缩编国外精品教材

Materials for Civil and Construction Engineers

土木工程材料

［美］Michael S. Mamlouk　　John P. Zaniewski　编

彭小芹　缩编

重庆大学出版社

Pearson Education, Inc.
Materials for Civil and Construction Engineers   ISBN:0-13-147714-5

Copyright © 2006 by Pearson Education, Inc.
Original language published by Pearson Education, Inc. All Rights reserved. 本书原版由美国培生出版集团出版。版权所有，盗印必究。

Chongqing University Press is authorized by Pearson Education, Inc to publish and distribute exclusively this English abridged edition. This edition is authorized for sale in the People's Republic of China only. Unauthorized export of this edition is a violation of the Copyright Act. No part of this publication may be reproduced or distributed by any means, or stored in a database or retrieval system, without the prior written permission of the publisher.

本书英文缩编由美国培生授权重庆大学出版社独家出版发行。此版本仅限在中国境内销售。未经授权的本书出口将被视为违反版权法的行为。未经出版者预先书面许可，不得以任何方式复制或发行本书的任何部分。

版贸核渝字(2007)第 82 号

图书在版编目(CIP)数据

土木工程材料/(美)马姆劳克(Mamlouk, M. S.)，
(美)赞尼尔斯基(Zaniewski, J. P.)编；彭小芹缩编．
—重庆：重庆大学出版社，2008.1(缩编国外精品教材)(2019.8 重印)
ISBN 978-7-5624-4312-4

Ⅰ.土⋯ Ⅱ.①马⋯②赞⋯③彭⋯ Ⅲ.土木工程—建筑材料—教材 Ⅳ.TU5

中国版本图书馆 CIP 数据核字(2007)第 176895 号

土木工程材料
Tumu Gongcheng Cailiao
[美] Michael S. Mamlouk   John P. Zaniewski
彭小芹　缩编

出　版　者：重庆大学出版社出版发行		地　　址：重庆市沙坪坝区大学城西路21号	
网　　址：http://www.cqup.com.cn		邮　　编：401331	
电　　话：(023)88617190　88617185(中小学)		传　　真：(023)88617186　88617166	
出 版 人：饶帮华			
责任编辑：贾兴文		版式设计：贾兴文	
责任校对：夏　宇		责任印制：张　策	

POD：重庆新生代彩印技术有限公司
发　行　者：全国新华书店经销
开　　本：787mm×1092mm　1/16　印张：20.5　字数：457 千
版　　次：2008 年 1 月第 1 版　2019 年 8 月第 3 次印刷
书　　号：ISBN 978-7-5624-4312-4
定　　价：48.00 元

缩编说明

《Materials for Civil and Construction Engineers》由 Michael S. Mamlouk 和 John P. Zaniewski 编写，2006 年由美国培生出版集团出版原书第二版。原书书名含义简单明了——土木工程师应该了解的材料，这一鲜明的观点也贯穿整本教材。为了与我国相关课程名称相符，缩编时将书名译为"土木工程材料"。原书在总体结构上与我国同类教材基本相同，从内容上分为 3 个部分，第一部分是材料工程导论；第二部分是用于土木工程的材料的特性；第三部分是评价材料性能的实验方法。每一章后面都附有习题。原书通篇从土木工程的角度诠释材料特性，深入浅出，紧密结合实际工程，有助于读者了解国外土木工程材料的应用技术和实验标准。适用于用作建筑材料专业和土木工程专业的双语教学和专业英语教学。

缩编作者根据国内土木工程专业基础课程"土木工程材料"或"建筑材料"的学时和教学要求，删除了原书第 2 章"Nature of Material"和其他有关章节以及部分与我国教学要求不相关的部分习题和附录，使本缩编教材的知识体系更加符合国内实际情况，结构更加紧凑。为便于读者阅读，书后附上部分较为生僻的专业词汇的英汉对照词汇表。

本书缩编过程中，陈科、刘芳、许国伟做了大量工作，特此致谢。

彭小芹

2007 年 11 月于重庆大学

PREFACE

A basic function of civil and construction engineering is to provide and maintain the infrastructure needs of society. The infrastructure includes buildings, water treatment and distribution systems, waste water removal and processing, dams, and highway and airport bridges and pavements. Although some civil and construction engineers are involved in the planning process, most are concerned with the design, construction, and maintenance of facilities. The common denominator among these responsibilities is the need to understand the behavior and performance of materials. Although not all civil and construction engineers need to be material specialists, a basic understanding of the material selection process, and the behavior of materials, is a fundamental requirement for all civil and construction engineers performing design, construction, and maintenance.

Material requirements in civil engineering and construction facilities are different from material requirements in other engineering disciplines. Frequently, civil engineering structures require tons of materials with relatively low replications of specific designs. Generally, the materials used in civil engineering have relatively low unit costs. In many cases, civil engineering structures are formed or fabricated in the field under adverse conditions. Finally, many civil engineering structures are directly exposed to detrimental effects of the environment.

The subject of engineering materials has advanced greatly in the last few decades. As a result, many of the conventional materials have either been replaced by more efficient materials or modified to improve their performance. Civil and construction engineers have to be aware of these advances and be able to select the most cost-effective material or use the appropriate modifier for the specific application at hand.

This text is organized into three parts: (1) introduction to materials engineering, (2) characteristics of materials used in civil and construction engineering, and (3) laboratory methods for the evaluation of materials.

The introduction to materials engineering includes information on the basic mechanistic properties of materials, environmental influences, and basic material classes. In addition, one of the responsibilities of civil and construction engineers is the inspection and quality control of materials in the construction process. This requires an understanding of material variability and testing procedures. The atomic structure of materials is covered in order to provide basic understanding of material behavior and to relate the molecular structure to the engineering response.

The second section, which represents a large portion of the book, presents the characteristics of the primary material types used in civil and construction engineering: steel, aluminum, concrete, masonry, asphalt, and wood. Since the discussion of concrete and asphalt materials requires a basic knowledge of aggregates, there is a chapter on aggregates. Moreover, since composites are gaining wide acceptance among engineers and are replacing many of the conventional materials, there is a chapter introducing composites.

The discussion of each type of material includes information on the following:

◆ Basic structure of the materials
◆ Material production process
◆ Mechanistic behavior of the material and other properties
◆ Environmental influences
◆ Construction considerations
◆ Special topics related to the material discussed in each chapter

Finally, each chapter includes an overview of various test procedures to introduce the test methods used with each material. However, the detailed description of the test procedures is left to the appropriate standards organizations such as the American Society for Testing and Materials (ASTM) and the American Association of State Highway and Transportation Officials (AASHTO). These ASTM and AASHTO standards are usually available in college libraries, and students are encouraged to use them. Also, there are sample problems in most chapters, as well as selected questions and problems at the end of each chapter. Answering these questions and problems will lead to a better understanding of the subject matter.

There are volumes of information available for each of these materials. It is not possible, or desirable, to cover these materials exhaustively in an introductory single text. Instead, this book limits the information to an introductory level, concentrates on current practices, and extracts information that is relevant to the general education of civil and construction engineers.

The content of the book is intended to be covered in one academic semester, although quarter system courses can definitely use it. The instructor of the course can also change the emphasis of some topics to match the specific curriculum of the department. Furthermore, since the course usually includes a laboratory portion, a number of laboratory test methods are described. The number of laboratory tests in the book is more than what is needed in a typical semester in order to provide more flexibility to the instructor to use the available equipment. Laboratory tests should be coordinated with the topics covered in the lectures so that the students get the most benefit from the laboratory experience.

The first edition of this textbook seemed to serve the needs of many universities and colleges. Therefore, the second edition is more of a refinement and updating of the book, with some notable additions. Several edits were made to the steel chapter to improve the description of heat treatments, phase diagram, and the heat-treating effects of welding. Also, a section on stainless steel was added, and current information on the structural uses of steel was provided. The cement and concrete chapters have been augmented with sections on hydration-control admixtures, recycled wash water, silica fume, self-consolidating concrete, and flowable fill. When the first edition was published, the Superpave mix design method was just being introduced to the industry. Now Superpave is a well-established method that has been field tested and revised to better meet the needs of the paving community. This development required a complete revision to the asphalt chapter to accommodate the current methods and procedures for both Performance Grading of asphalt binders and the Superpave mix design method. The chapter on wood was revised to provide information on recent manufactured wood products that became available in the last several years. Also, since fiber reinforced polymer composites have been more commonly used in retrofitting old and partially damaged structures, several examples were added in the chapter on composites. In the laboratory manual, an experiment on dry-rodded unit weight of aggregate that is used in portland cement concrete (PCC) proportioning was added and the experiment on creep of asphalt concrete was deleted for lack of use.

In addition to the technical content revisions, there are over 100 new figures to display concepts and equipment. Multiple sample problems and homework problems have been added to each chapter to allow professors to vary assignments between semesters.

The authors would like to acknowledge the contributions of Drs. Barzin Mobasher and Chris Lawrence of Arizona State University, Mr. Jim Willson and Mr. Paul Mueller of the Portland Cement Association, Dr. Mansour Solimanian of Pennsylvania State University, Mr. Lary Lenke of the University of New Mexico, and Dr. Nabil Grace of Lawrence Tech University for their advice and for providing some photos and homework problems. Appreciation also goes to Mr. Sherif El-Badawy of Arizona State University for his contribution in the preparation of the solutions manual.

CONTENTS

Chapter 1 Materials Engineering Concepts 1
1.1 Economic Factors 2
1.2 Mechanical Properties 4
   1.2.1 Loading Conditions 4
   1.2.2 Stress-Strain Relations 5
   1.2.3 Elastic Behavior 6
   1.2.4 Elastoplastic Behavior 9
   1.2.5 Work and Energy 12
   1.2.6 Time-Dependent Response 13
   1.2.7 Rheological Models 15
   1.2.8 Temperature and Time Effects 19
   1.2.9 Failure and Safety 20
1.3 Nonmechanical Properties 22
   1.3.1 Density and Unit Weight 22
   1.3.2 Thermal Expansion 23
   1.3.3 Surface Characteristics 25
1.4 Production and Construction 26
1.5 Aesthetic Characteristics 26
1.6 Material Variability 27
   1.6.1 Sampling 28
   1.6.2 Normal Distribution 29
   1.6.3 Control Charts 29
   1.6.4 Experimental Error 32
1.7 Laboratory Measuring Devices 32
   1.7.1 Dial Gauge 33
   1.7.2 Linear Variable Differential Transformer (LVDT) 34
   1.7.3 Strain Gauge 36
   1.7.4 Proving Ring 37
   1.7.5 Load Cell 38
Summary 39
Questions and Problems 40

2   **Materials for Civil and Construction Engineers**

    References ··· 42

Chapter 2   Steel ··· 43
2.1   Steel Production ··· 44
2.2   Mechanical Testing of Steel ·· 46
   2.2.1   Tension Test ··· 46
   2.2.2   Torsion Test ·· 53
   2.2.3   Charpy V Notch Impact Test ······································ 55
   2.2.4   Bend Test ·· 57
   2.2.5   Hardness Test ··· 58
   2.2.6   Ultrasonic Testing ·· 59
2.3   Steel Corrosion ·· 59
   Summary ·· 61
   Questions and Problems ·· 61
   References ·· 61

Chapter 3   Aluminum ·· 63
3.1   Aluminum Production ··· 64
3.2   Aluminum Testing and Properties ·· 67
3.3   Corrosion ·· 73
   Summary ·· 73
   Questions and Problems ·· 74
   References ·· 74

Chapter 4   Aggregates ··· 75
4.1   Aggregate Sources ··· 75
4.2   Geological Classification ·· 76
4.3   Evaluation of Aggregate Sources ··· 76
4.4   Aggregate Uses ·· 77
4.5   Aggregate Properties ··· 78
   4.5.1   Particle Shape and Surface Texture ··························· 79
   4.5.2   Soundness and Durability ··· 81
   4.5.3   Toughness, Hardness, and Abrasion Resistance ······· 82
   4.5.4   Absorption ·· 83
   4.5.5   Specific Gravity ··· 84
   4.5.6   Bulk Unit Weight and Voids in Aggregate ··············· 86

4.5.7	Strength and Modulus	87
4.5.8	Gradation and Maximum Size	88
4.5.9	Deleterious Substances in Aggregate	104
4.5.10	Alkali-Aggregate Reactivity	104
4.5.11	Affinity for Asphalt	105

4.6 Sampling Aggregates ··· 106
Summary ··· 107
Questions and Problems ··· 107
References ··· 108

Chapter 5 Portland Cement ··· 110
5.1 Portland Cement Production ··· 110
5.2 Chemical Composition of Portland Cement ··· 112
5.3 Fineness of Portland Cement ··· 113
5.4 Specific Gravity of Portland Cement ··· 113
5.5 Hydration of Portland Cement ··· 114
   5.5.1 Structure Development in Cement Paste ··· 115
   5.5.2 Evaluation of Hydration Progress ··· 116
5.6 Voids in Hydrated Cement ··· 117
5.7 Properties of Hydrated Cement ··· 117
   5.7.1 Setting ··· 118
   5.7.2 Soundness ··· 119
   5.7.3 Compressive Strength ··· 120
5.8 Water-Cementitious Materials Ratio ··· 120
5.9 Types of Portland Cement ··· 121
   5.9.1 Standard Portland Cement Types ··· 121
   5.9.2 Other Cement Types ··· 124
5.10 Mixing Water ··· 125
   5.10.1 Acceptable Criteria ··· 125
   5.10.2 Disposal and Reuse of Concrete Wash Water ··· 127
5.11 Admixtures for Concrete ··· 127
   5.11.1 Air Entrainers ··· 128
   5.11.2 Water Reducers ··· 130
   5.11.3 Retarders ··· 133
   5.11.4 Hydration-Control Admixtures ··· 133
   5.11.5 Accelerators ··· 133

5.11.6　Supplementary Cementitious Admixtures　135
5.11.7　Specialty Admixtures　138
Summary　139
Questions and Problems　139
References　140

Chapter 6　Portland Cement Concrete　141

6.1　Proportioning of Concrete Mixes　142
　6.1.1　Basic Steps for Weight and Absolute Volume Methods　143
　6.1.2　Mixing Concrete for Small Jobs　158
6.2　Mixing, Placing, and Handling Fresh Concrete　160
　6.2.1　Ready-Mixed Concrete　161
　6.2.2　Mobile Batcher Mixed Concrete　161
　6.2.3　Depositing Concrete　161
　6.2.4　Pumped Concrete　161
　6.2.5　Vibration of Concrete　162
　6.2.6　Pitfalls and Precautions for Mixing Water　162
　6.2.7　Measuring Air Content in Fresh Concrete　162
　6.2.8　Spreading and Finishing Concrete　164
6.3　Curing Concrete　164
　6.3.1　Ponding or Immersion　166
　6.3.2　Spraying or Fogging　166
　6.3.3　Wet Coverings　166
　6.3.4　Impervious Papers or Plastic Sheets　166
　6.3.5　Membrane-Forming Compounds　167
　6.3.6　Forms Left in Place　167
　6.3.7　Steam Curing　167
　6.3.8　Insulating Blankets or Covers　167
　6.3.9　Electrical, Hot Oil, and Infrared Curing　168
　6.3.10　Curing Period　168
6.4　Properties of Hardened Concrete　168
　6.4.1　Early Volume Change　168
　6.4.2　Creep Properties　169
　6.4.3　Permeability　169
　6.4.4　Stress-Strain Relationship　170
6.5　Testing of Hardened Concrete　172

    6.5.1 Compressive Strength Test 173
    6.5.2 Split-Tension Test 174
    6.5.3 Flexure Strength Test 175
    6.5.4 Rebound Hammer Test 176
    6.5.5 Penetration Resistance Test 177
    6.5.6 Ultrasonic Pulse Velocity Test 178
    6.5.7 Maturity Test 179
6.6 Alternatives to Conventional Concrete 179
    6.6.1 Self-Consolidating Concrete 179
    6.6.2 Flowable Fill 181
    6.6.3 Shotcrete 182
    6.6.4 Lightweight Concrete 182
    6.6.5 Heavyweight Concrete 183
    6.6.6 High-Strength Concrete 183
    6.6.7 Shrinkage-Compensating Concrete 184
    6.6.8 Polymers and Concrete 184
    6.6.9 Fiber-Reinforced Concrete 184
    6.6.10 Roller-Compacted Concrete 185
    6.6.11 High-Performance Concrete 186
Summary 187
Questions and Problems 188
References 189

Chapter 7 Masonry 191

7.1 Masonry Units 191
    7.1.1 Concrete Masonry Units 192
    7.1.2 Clay Bricks 196
7.2 Mortar 199
7.3 Grout 200
7.4 Plaster 200
Summary 200
Questions and Problems 200
References 202

Chapter 8 Asphalt Binders and Asphalt Mixtures 203

8.1 Types of Asphalt Products 206

8.2　Uses of Asphalt ... 207
8.3　Temperature Susceptibility of Asphalt ... 208
8.4　Chemical Properties of Asphalt ... 210
8.5　Superpave and Performance Grade Binders ... 211
8.6　Asphalt Concrete ... 212
8.7　Asphalt Concrete Mix Design ... 213
　8.7.1　Specimen Preparation in the Laboratory ... 213
　8.7.2　Density and Voids Analysis ... 215
　8.7.3　Superpave Mix Design ... 219
　8.7.4　Superpave Simple Performance Tests (SPT) ... 227
　8.7.5　Marshall Method of Mix Design ... 229
　8.7.6　Hveem Method of Mix Design ... 236
　8.7.7　Evaluation of Moisture Susceptibility ... 238
8.8　Characterization of Asphalt Concrete ... 239
　8.8.1　Indirect Tensile Strength ... 240
　8.8.2　Diametral Tensile Resilient Modulus ... 241
　8.8.3　Freeze and Thaw Test ... 242
　8.8.4　Use of Rheological Models to Analyze Time-Dependent Response ... 243
8.9　Asphalt Concrete Production ... 243
8.10　Recycling of Asphalt Concrete ... 243
　8.10.1　Surface Recycling ... 244
　8.10.2　Central Plant Recycling ... 244
　8.10.3　In-Place Recycling ... 245
8.11　Additives ... 245
　8.11.1　Fillers ... 245
　8.11.2　Extenders ... 245
　8.11.3　Rubber ... 245
　8.11.4　Plastics ... 246
　8.11.5　Antistripping Agents ... 246
　8.11.6　Others ... 246
Summary ... 246
Questions and Problems ... 247
References ... 247

Chapter 9　Wood ... 249
9.1　Structure of Wood ... 251

    9.1.1 Growth Rings ... 251
    9.1.2 Anisotropic Nature of Wood ... 252
  9.2 Chemical Composition ... 254
  9.3 Physical Properties ... 254
    9.3.1 Specific Gravity and Density ... 255
    9.3.2 Thermal Properties ... 255
    9.3.3 Electrical Properties ... 256
  9.4 Mechanical Properties ... 257
    9.4.1 Modulus of Elasticity ... 257
    9.4.2 Strength Properties ... 257
    9.4.3 Creep ... 258
    9.4.4 Damping Capacity ... 258
  9.5 Testing to Determine Mechanical Properties ... 258
    9.5.1 Static Bending Test ... 260
    9.5.2 Compression Tests ... 261
  Summary ... 261
  Questions and Problems ... 262
  References ... 263

Appendix ... 264
Laboratory Manual ... 264
1. Introduction to Measuring Devices ... 265
2. Tension Test of Steel and Aluminum ... 268
3. Impact Test of Steel ... 272
4. Sieve Analysis of Aggregates ... 274
5. Specific Gravity and Absorption of Coarse Aggregate ... 277
6. Specific Gravity and Absorption of Fine Aggregate ... 279
7. Bulk Unit Weight and Voids in Aggregate ... 282
8. Slump of Freshly Mixed Portland Cement Concrete ... 284
9. Unit Weight and Yield of Freshly Mixed Concrete ... 285
10. Air Content of Freshly Mixed Concrete by Pressure Method ... 287
11. Air Content of Freshly Mixed Concrete by Volumetric Method ... 288
12. Making and Curing Concrete Cylinders and Beams ... 290
13. Flexural Strength of Concrete ... 292
14. Penetration Test of Asphalt Cement ... 294
15. Absolute Viscosity Test of Asphalt ... 296

16. Preparing and Determining the Density of Hot-Mix Asphalt (HMA) Specimens by Means of the Superpave Gyratory Compactor ·· 297
17. Preparation of Asphalt Concrete Specimens Using the Marshall Compactor ············ 299
18. Bulk Specific Gravity of Compacted Bituminious Mixtures ································ 301
19. Marshall Stability and Flow of Asphalt Concrete ·· 302
20. Bending and Compression Tests of Wood ·· 304
21. Tensile Properties of Plastics ·· 308

Glossary ·· 311

CHAPTER 1

MATERIALS ENGINEERING CONCEPTS

Materials engineers are responsible for the selection, specification, and quality control of materials to be used in a job. These materials must meet certain classes of criteria or materials properties (Ashby and Jones 1980). These classes of criteria include:

- economic factors
- mechanical properties
- nonmechanical properties
- production/construction considerations
- aesthetic properties

When engineers select the material for a specific application, they must consider the various criteria and make compromises. Both the client and the purpose of the facility or structure dictate, to a certain extent, the emphasis that will be placed on the different criteria.

Civil and construction engineers must be familiar with materials used in the construction of a wide range of structures. Materials most frequently used include steel, aggregate, concrete, masonry, asphalt, and wood. Materials used to a lesser extent include aluminum, glass, plastics and fiber-reinforced composites. Geotechnical engineers make a reasonable case for including soil as the most widely used engineering material, since it provides the basic support for all civil engineering structures. However, the properties of soils will not be discussed in this text, because this is generally the topic of a separate course.

Recent advances in the technology of civil engineering materials have resulted in the development of better quality, more economical and safer materials. These materials are commonly referred to as high-performance materials. Because more is known about the molecular structure of materials and because of the continuous research efforts by scientists and engineers, new materials such as polymers, adhesives, composites, geotextiles, coatings, cold-formed metals, and various synthetic products are competing with traditional civil engineering materials. In addition, improvements have been made to existing materials by changing their molecular structures or including additives to improve quality, economy and performance. For example, superplasticizers have made a breakthrough in the concrete industry, allowing the production of much stronger concrete. Joints made of elastomeric materials have improved the safety of high-rise structures in earthquake-active areas. Lightweight synthetic aggregates have decreased the weight of concrete structures, allowing small cross-sectional areas of components. Polymers have been mixed with asphalt, allowing pavements to last longer under the effect of vehicle loads and environmental conditions.

The field of fiber composite materials has developed rapidly in the last 30 years. Many recent civil engineering projects have used fiber-reinforced composites. These advanced composites compete with traditional materials due to their higher strength-to-weight ratio and their ability to overcome such short-comings as corrosion. For example, fiber-reinforced concrete has much greater toughness than conventional portland cement concrete. Composites can replace reinforcing steel in concrete structures. In fact, composites have allowed the construction of structures that could not have been built in the past.

The nature and behavior of civil engineering materials are as complicated as those of materials used in any other field of engineering. Due to the high quantity of materials used in civil engineering projects, the civil engineer frequently works with locally available materials that are not as highly refined as the materials used in other engineering fields. As a result, civil engineering materials frequently have highly variable properties and characteristics.

This chapter reviews the manner in which the properties of materials affect their selection and performance in civil engineering applications. In addition, the chapter reviews some basic definitions and concepts of engineering mechanics required for understanding material behavior. The variable nature of material properties is also discussed so that the engineer will understand the concepts of precision and accuracy, sampling, quality assurance, and quality control. Finally, instruments used for measuring material response are described.

1.1 Economic Factors

The economics of the material selection process are affected by much more than just the cost of the material. Factors that should be considered in the selection of the material include:

- availability and cost of raw materials
- manufacturing costs
- transportation
- placing
- maintenance

The materials used for civil engineering structures have changed over time. Early structures were constructed of stone and wood. These materials were in ready supply and could be cut and shaped with available tools. Later, cast iron was used, because mills were capable of crudely refining iron ore. As the industrial revolution took hold, quality steel could be produced in the quantities required for large structures. In addition, portland cement, developed in the mid-1800s, provided civil engineers with a durable inexpensive material with broad applications.

Due to the efficient transportation system in the United States, availability is not as much of an issue as it once was in the selection of a material. However, transportation can significantly add to the cost of the materials at the job site. For example, in many locations in the United States, quality aggregates for concrete and asphalt are in short supply. The closest aggregate source to Houston, Texas, is 150 km (90 miles) from the city. This haul distance approximately doubles the cost of the aggregates in the city and hence puts concrete at a disadvantage compared with steel.

The type of material selected for a job can greatly affect the ease of construction and the construction costs and time. For example, the structural members of a steel-frame building can be fabricated in a shop, transported to the job site, lifted into place with a crane and bolted or welded together. In contrast, for a reinforced concrete building, the forms must be built; reinforcing steel placed; concrete mixed, placed, and allowed to cure; and the forms removed. Constructing the concrete frame building can be more complicated and time consuming than constructing steel structures. To overcome this shortcoming, precast concrete units commonly have been used, especially for bridge construction.

All materials deteriorate over time and with use. This deterioration affects both the maintenance cost and the useful life of the structure. The rate of deterioration varies among materials. Thus, in analyzing the economic selection of a material, the life cycle cost should be evaluated in addition to the initial costs of the structure.

1.2 Mechanical Properties

The mechanical behavior of materials is the response of the material to external loads. All materials deform in response to loads; however, the specific response of a material depends on its properties, the magnitude and type of load, and the geometry of the element. Whether the material "fails" under the load conditions depends on the failure criterion. Catastrophic failure of a structural member, resulting in the collapse of the structure, is an obvious material failure. However, in some cases the failure is more subtle, but with equally severe consequences. For example, pavement may fail due to excessive roughness at the surface, even though the stress levels are well within the capabilities of the material. A building may have to be closed due to excessive vibrations by wind or other live loads, although it could be structurally sound. These are examples of functional failures.

1.2.1 Loading Conditions

One of the considerations in the design of a project is the type of loading the structure will be subjected to during its design life. The two basic types of loads are static and dynamic. Each type affects the material differently, and frequently the interactions between the load types are important. Civil engineers encounter both when designing a structure.

Static loading implies a sustained loading of the structure over a period of time. Generally, static loads are slowly applied such that no shock or vibration is generated in the structure. Once applied, the static load may remain in place or be removed slowly. Loads that remain in place for an extended period of time are called sustained (dead) loads. In civil engineering, much of the load the materials must carry is due to the weight of the structure and equipment in the structure.

Loads that generate a shock or vibration in the structure are dynamic loads. Dynamic loads can be classified as periodic, random, or transient, as shown in Figure 1.1 (Richart et al. 1970). A periodic load, such as a harmonic or sinusoidal load, repeats itself with time. For example, rotating equipment in a building can produce a vibratory load. In a random load, the load pattern never repeats, such as that produced by earthquakes. Transient load, on the other hand, is an impulse load that is applied over a short time interval, after which the vibrations decay until the system returns to a rest condition. For example, bridges must be designed to withstand the transient loads of trucks.

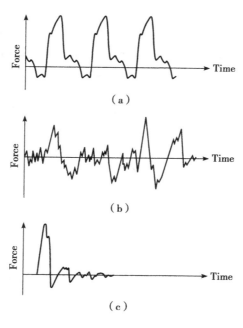

Figure 1.1 Types of dynamic loads: (a) periodic, (b) random, and (c) transient.

1.2.2 Stress-Strain Relations

Materials deform in response to loads or forces. In 1678, Robert Hooke published the first findings that documented a linear relationship between the amount of force applied to a member and its deformation. The amount of deformation is proportional to the properties of the material and its dimensions. The effect of the dimensions can be normalized. Dividing the force by the cross-sectional area of the specimen normalizes the effect of the loaded area. The force per unit area is defined as the stress σ in the specimen (i. e. , σ=Force/Area). Dividing the deformation by the original length is defined as strain ε of the specimen (i. e. , ε=Change in length/Original length). Much useful information about the material can be determined by plotting the stress-strain diagram.

Figure 1.2 shows typical uniaxial tensile or compressive stress-strain curves for several engineering materials. Figure 1.2(a) shows a linear stress-strain relationship up to the point where the material fails. Glass and chalk are typical of materials exhibiting this tensile behavior. Figure 1.2(b) shows the behavior of steel in tension. Here, a linear relationship is obtained up to a certain point (proportional limit), after which the material deforms without much increase in stress. On the other hand, aluminum alloys in tension exhibit a linear stress-strain relation up to the proportional limit, after which a nonlinear relation follows, as illustrated in Figure 1.2(c). Figure 1.2(d) shows a nonlinear relation throughout the whole range. Concrete and other materials exhibit this relationship, although the first portion of the curve for concrete is very close to being linear. Soft rubber in tension differs from most materials in such a way that it shows an almost linear stress-strain relationship followed by a reverse curve, as shown in Figure 1.2(e).

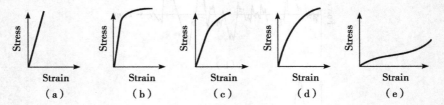

Figure 1.2 Typical uniaxial stress-strain diagrams for some engineering materials: (a) glass and chalk, (b) steel, (c) aluminum alloys, (d) concrete, and (e) soft rubber.

1.2.3 Elastic Behavior

If a material exhibits true elastic behavior, it must have an instantaneous response (deformation) to load, and the material must return to its original shape when the load is removed. Many materials, including most metals, exhibit elastic behavior, at least at low stress levels. As will be discussed in Chapter 2, elastic deformation does not change the arrangement of atoms within the material, but rather it stretches the bonds between atoms. When the load is removed the atomic bonds return to their original position.

Young observed that different elastic materials have different proportional constants between stress and strain. For a homogeneous, isotropic, and linear elastic material, the proportional constant between normal stress and normal strain of an axially loaded member is the modulus of elasticity or Young's modulus, E, and is equal to

$$E = \frac{\sigma}{\varepsilon} \tag{1.1}$$

where σ is the normal stress and ε is the normal strain.

In the axial tension test, as the material is elongated, there is a reduction of the cross section in the lateral direction. In the axial compression test, the opposite is true. The ratio of the lateral strain, ε_l, to the axial strain, ε_a, is Poisson's ratio,

$$\upsilon = \frac{-\varepsilon_l}{\varepsilon_a} \tag{1.2}$$

Since the axial and lateral strains will always have different signs, the negative sign is used in Equation 1.2 to make the ratio positive. Poisson's ratio has a theoretical range of 0.0 to 0.5, where 0.0 is for a compressible material in which the axial and lateral directions are not affected by each other. The 0.5 value is for a material that does not change its volume when the load is applied. Most solids have Poisson's ratios between 0.10 and 0.45.

Although Young's modulus and Poisson's ratio were defined for the uniaxial stress condition, they are important when describing the three-dimensional stress-strain relationships, as well. If a homogeneous, isotropic cubical element with linear elastic response is subjected to normal stresses σ_x, σ_y, and σ_z in the three orthogonal directions (as shown in Figure 1.3), the normal strains ε_x, ε_y, and ε_z can be computed by the generalized Hooke's law,

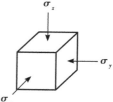

Figure 1.3 Normal stresses applied on a cubical element.

$$\varepsilon_x = \frac{\sigma_x - \upsilon(\sigma_y + \sigma_z)}{E}$$

$$\varepsilon_y = \frac{\sigma_y - \upsilon(\sigma_z + \sigma_x)}{E} \quad (1.3)$$

$$\varepsilon_z = \frac{\sigma_z - \upsilon(\sigma_x + \sigma_y)}{E}$$

Sample Problem 1.1

A cube made of an alloy with dimensions of 50 mm×50 mm×50 mm is placed into a pressure chamber and subjected to a pressure of 90 MPa. If the modulus of elasticity of the alloy is 100 GPa and Poisson's ratio is 0.28, what will be the length of each side of the cube, assuming that the material remains within the elastic region?

Solution

$$\varepsilon_x = [\sigma_x - \nu(\sigma_y + \sigma_z)]/E = [-90 - 0.28 \times (-90 - 90)]/100,000$$
$$= -0.000396 \text{ m/m}$$
$$\varepsilon_y = \varepsilon_z = -0.000396 \text{ m/m}$$
$$\Delta x = \Delta y = \Delta z = -0.000396 \times 50 = -0.0198 \text{ mm}$$
$$L_{new} = 50 - 0.0198 = 49.9802 \text{ mm}$$

Linearity and elasticity should not be confused. A linear material's stress-strain relation follows a straight line. An elastic material returns to its original shape when the load is removed and reacts instantaneously to changes in load. For example, Figure 1.4(a) represents a linear elastic behavior, while Figure 1.4(b) represents a nonlinear elastic behavior.

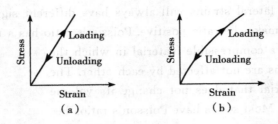

Figure 1.4 Elastic behavior: (a) linear and (b) nonlinear.

For materials that do not display any linear behavior, such as concrete and soils, determining a Young's modulus or elastic modulus can be problematical. There are several options for arbitrarily defining the modulus for these materials. Figure 1.5 shows four options: the initial tangent, tangent, secant and chord moduli. The initial tangent modulus is the slope of the tangent of the stress-strain curve at the origin. The tangent modulus is the slope of the tangent at a point on the stress-strain curve. The secant modulus is the slope of a chord drawn between the origin and an arbitrary point on the stress-strain curve. The chord modulus is the slope of a chord drawn between two points on the stress-strain curve. The selection of which modulus to use for a nonlinear material depends on the stress or strain level at which the material typically is used. Also, when determining the tangent, secant, or chord modulus, the stress or strain levels must be defined.

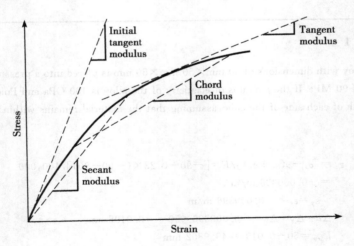

Figure 1.5 Methods for approximating modulus.

Table 1.1 shows typical modulus and Poisson's ratio values for some materials at room temperature. Note that some materials have a range of modulus values rather than a distinct value. Several factors affect the modulus, such as curing level and proportions of components of concrete or the direction of loading relative to the grain of wood.

Table 1.1 Typical Modulus and Poisson's Ratio Values (Room Temperature)

Material	Modulus GPa (psi×10⁶)	Poisson's Ratio
Aluminum	69-75(10-11)	0.33
Brick	10-17(1.5-2.5)	0.23-0.40
Cast iron	75-169(11-23)	0.17
Concrete	14-40(2-6)	0.11-0.21
Copper	110(16)	0.35
Epoxy	3-140(0.4-20)	
Glass	62-70(9-10)	0.25
Limestone	58(8.4)	
Rubber(soft)	0.001-0.014(0.00015-0.002)	0.49
Steel	207(30)	0.27
Tungsten	407(59)	0.28
Wood	6-15(0.9-2.2)	

1.2.4 Elastoplastic Behavior

For some materials, as the stress applied on the specimen is increased, the strain will proportionally increase up to a point; after this point the strain will increase with little additional stress. In this case, the material exhibits linear elastic behavior followed by plastic response. The stress level at which the behavior changes from elastic to plastic is the elastic limit. When the load is removed from the specimen, some of the deformation will be recovered and some of the deformation will remain as seen in Figure 1.6(a). As discussed in Chapter 2, plastic behavior indicates permanent deformation of the specimen so that it does not return to its original shape when the load is removed. This indicates that when the load is applied, the atomic bonds stretch, creating an elastic response; then the atoms actually slip relative to each other. When the load is removed, the atomic slip does not recover; only the atomic stretch is recovered (Callister 2003).

Several models are used to represent the behavior of materials that exhibit both elastic and plastic responses. Figure 1.6(b) shows a linear elastic-perfectly plastic response in which the material exhibits a linear elastic response upon loading, followed by a completely plastic response. If such material is unloaded after it has plasticly deformed, it will rebound in a linear elastic manner and will follow a straight line parallel to the elastic portion, while some permanent deformation will remain. If the material is loaded again, it will have a linear elastic response followed by plastic response at the same level of stress at which the material was unloaded (Popov 1968).

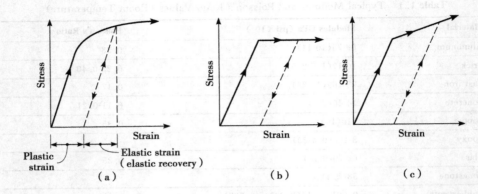

Figure 1.6 Stress-strain behavior of plastic materials: (a) example of loading and unloading, (b) elastic-perfectly plastic, and (c) elastoplastic with strain hardening.

Figure 1.6(c) shows an elastoplastic response in which the first portion is an elastic response followed by a combined elastic and plastic response. If the load is removed after the plastic deformation, the stress-strain relation will follow a straight line parallel to the elastic portion; consequently, some of the strain in the material will be removed, and the remainder of the strain will be permanent. Upon reloading, the material again behaves in a linear elastic manner up to the stress level that was attained in the previous stress cycle. After that point the material will follow the original stress-strain curve. Thus, the stress required to cause plastic deformation actually increases. This process is called strain hardening or work hardening. Strain hardening is beneficial in some cases, since it allows more stress to be applied without permanent deformation. Mild steel is an example of material that experiences strain hardening during plastic deformation.

Some materials exhibit strain softening, in which plastic deformation causes weakening of the material. Portland cement concrete is a good example of such a material. In this case, plastic deformation causes microcracks at the interface between aggregate and cement paste.

Sample Problem 1.2

An elastoplastic material with strain hardening has the stress-strain relation shown in Figure 1.6(c). The modulus of elasticity is 25×10^6 psi, yield strength is 70 ksi, and the slope of the strain-hardening portion of the stress-strain diagram is 3×10^6 psi.

    a. Calculate the strain that corresponds to a stress of 80 ksi.
    b. If the 80-ksi stress is removed, calculate the permanent strain.

Solution

(a) $\varepsilon = (70{,}000/25 \times 10^6) + [(80{,}000 - 70{,}000)/(3 \times 10^6)] = 0.0028 + 0.0033$

$= 0.0061$ in./in.

(b) $\varepsilon_{permanent} = 0.0061 - [80,000/(25 \times 10^6)] = 0.0061 - 0.0032$
$= 0.0029$ in./in.

Materials that do not undergo plastic deformation prior to failure, such as concrete, are said to be brittle, whereas materials that display appreciable plastic deformation, such as mild steel, are ductile. Generally, ductile materials are preferred for construction. When a brittle material fails, the structure can collapse in a catastrophic manner. On the other hand, overloading a ductile material will result in distortions of the structure, but the structure will not necessarily collapse. Thus, the ductile material provides the designer with a margin of safety.

Figure 1.7(a) demonstrates three concepts of the stress-strain behavior of elastoplastic materials. The lowest point shown on the diagram is the proportional limit, defined as the transition point between linear and nonlinear behavior. The second point is the elastic limit, which is the transition between elastic and plastic behavior. However, most materials do not display an abrupt change in behavior from elastic to plastic. Rather, there is a gradual, almost imperceptible transition between the behaviors, making it difficult to locate an exact transition point (Polowski and Ripling 1966). For this reason, arbitrary methods such as the offset and the extension methods, are used to identify the elastic limit, thereby defining the yield stress (yield strength). In the offset method, a specified offset is measured on the abscissa, and a line with a slope equal to the initial tangent modulus is drawn through this point. The point where this line intersects the stress-strain curve is the offset yield stress of the material, as seen in Figure 1.7(a). Different offsets are used for different materials (Table 1.2). The extension yield *stress* is located where a vertical projection, at a specified strain level, intersects the stress-strain curve. Figure 1.7(b) shows the yield stress corresponding to 0.5% extension.

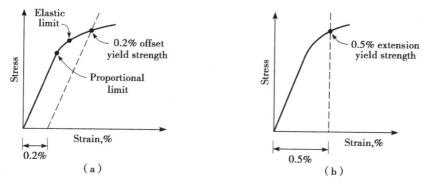

Figure 1.7 Methods for estimating yield stress: (a) offset method and (b) extension method.

Table 1.2 Offset Values Typically Used to Determine Yield Stress

Material	Stress Condition	Offset(%)	Corresponding Strain
Steel	Tension	0.20	0.0020
Wood	Compression parallel to grain	0.05	0.0005
Gray cast iron	Tension	0.05	0.0005
Concrete	Compression	0.02	0.0002
Aluminum alloys	Tension	0.20	0.0020
Brass and bronze	Tension	0.35	0.0035

Sample Problem 1.3

A rod made of aluminum alloy, with a gauge length of 100 mm, diameter of 10 mm, and yield strength of 150 MPa, was subjected to a tensile load of 5.85 kN. If the gauge length was changed to 100.1 mm and the diameter was changed to 9.9967 mm, calculate the modulus of elasticity and Poisson's ratio.

Solution

$$\sigma = P/A = (5850 \text{ N})/[\pi(5\times 10^{-3} \text{ m})^2] = 74.5\times 10^6 \text{ Pa} = 74.5 \text{ MPa}$$

Since the applied stress is well below the yield strength, the material is within the elastic region.

$$\varepsilon_a = \Delta L/L = (100.1-100)/100 = 0.001$$
$$E = \sigma/\varepsilon_a = 74.5/0.001 = 74,500 \text{ MPa} = 74.5 \text{ GPa}$$
$$\varepsilon_l = \text{change in diameter/diameter} = (9.9967-10)/10 = -0.00033$$
$$\nu = -\varepsilon_l/\varepsilon_a = 0.00033/0.001 = 0.33$$

1.2.5 Work and Energy

When a material is tested, the testing machine is actually generating a force in order to move or deform the specimen. Since work is force times distance, the area under a force-displacement curve is the work done on the specimen. When the force is divided by the cross-sectional area of the specimen to compute the stress, and the deformation is divided by the length of the specimen to compute the strain, the force-displacement diagram becomes a stress-strain diagram. However, the area under the stress-strain diagram no longer has the units of work. By manipulating the units of the stress-strain diagram, we can see that the area under the stress-strain diagram equals the work per unit volume of material required to deform or fracture the material. This is a useful concept, for it tells us the energy that is required to deform or fracture the material. Such information is used for selecting materials to use where energy must be absorbed by the member. The area under the elastic portion of the curve is the modulus of resilience [Figure 1.8(a)]. The amount of energy required to fracture a specimen is a measure of the toughness of the material, as in Figure 1.8(b). As shown in Figure 1.8(c), a high-strength material is not necessarily a tough material. For instance, as will be discussed in Chapter 3, increasing the carbon content of steel increases the yield strength, but reduces ductility. Therefore, the strength is increased, but the toughness may be reduced.

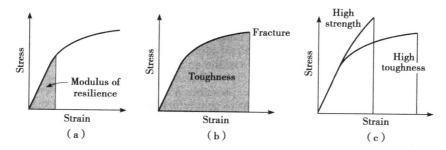

Figure1.8 Areas under stress-strain curves: (a) modulus of resilience, (b) toughness, and (c) high-strength and high-toughness materials.

1.2.6 Time-Dependent Response

The previous discussion assumed that the strain was an immediate response to stress. This is an assumption for elastic materials. However, no material has this property under all conditions. In some cases, materials have a delayed response. The amount of deformation depends on the duration of the load, the temperature, and the material characteristics. There are several mechanisms associated with time-dependent deformation, such as creep and viscous flow. There is no clear distinction between these terms. Creep is generally associated with long-term deformations and can occur in metals, ionic and covalent crystals, and amorphous materials. On the other hand, viscous flow is associated only with amorphous materials and can occur under short-term load duration. For example, concrete, a material with predominantly covalent crystals, can creep over a period of decades. Asphalt concrete pavements, an amorphous-binder material, can have ruts caused by the accumulated effect of viscous flows resulting from traffic loads with a load duration of only a fraction of a second.

Creep of metals is not relevant in typical civil engineering applications. In steel, creep can occur at temperatures greater than 30% of the melting point on the absolute scale. This may be a concern in the design of boilers and nuclear reactor containment vessels. Creep of metals occurs in three phases. The first phase is the result of dislocation movements in the molecular structure of the metal. The second phase is associated with slip at the grain boundaries, similar to plastic deformation, but accelerated due to the high temperature. The third phase is associated with an increase in the strain due to a reduction of the cross section of the specimen. Creep is also considered in the design of wood and advanced composite structural members. Wood elements loaded for a few days can carry higher stresses than elements designed to carry "permanent" loads. On the other hand, creep of concrete is associated with microcracking at the interface of the cement paste and the aggregate particles (Mehta and Monteiro 1993).

The viscous flow models are similar in nature to Hooke's law. In linearly viscous materials, the rate of deformation is proportional to the stress level. These materials are not compressible and do not recover when the load is removed. Materials with these characteristics are Newtonian fluids.

Figure 1.9(a) shows a typical creep test in which a constant compressive stress is applied to an asphalt concrete specimen. In this case, an elastic strain will develop, followed by time-dependent strain or creep. If the specimen is unloaded, a part of the strain will recover instantaneously, while the remaining strain will recover, either completely or partially, over a period of time. Another phenomenon typical of time-dependent materials is relaxation, or dissipation of stresses with time. For example, if an asphalt concrete specimen is placed in a loading machine and subjected to a constant strain, the stress within the specimen will initially be high, then gradually dissipate due to relaxation as shown in Figure 1.9(b). Relaxation is an important concern in the selection of steel for a prestressed concrete design.

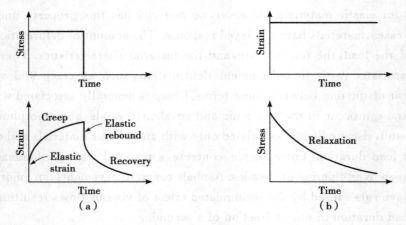

Figure 1.9 Behavior of time-dependent materials: (a) creep and (b) relaxation.

Materials exhibiting both viscous and elastic responses are known as viscoelastic. In viscoelasticity, there are two approaches used to describe how stresses, strains, and time are interrelated. One approach is to postulate mathematical relations between these parameters based on material functions obtained from laboratory tests. The other approach is based on combining a number of discrete rheological elements to form rheological models, which describe the material response.

1.2.7 Rheological Models

Rheological models are used to model mechanically the time-dependent behavior of materials. There are many different modes of material deformation, particularly in polymer materials. These materials cannot be described as simply elastic, viscous, etc. However, these materials can be modeled by a combination of simple physical elements. The simple physical elements have characteristics that can be easily visualized. Rheology uses three basic elements, combined in either series or parallel to form models that define complex material behaviors. The three basic rheological elements, Hookean, Newtonian, and St. Venant, are shown in Figure 1.10 (Polowski and Ripling 1966).

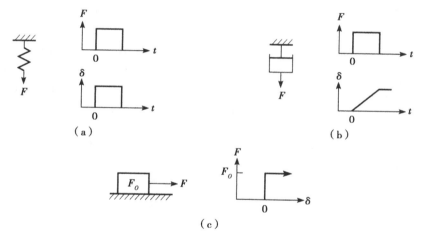

Figure 1.10 Basic elements used in rheology: (a) Hookean, (b) Newtonian, and (c) St. Venant.

The Hookean element, as in Figure 1.10(a), has the characteristics of a linear spring. The deformation δ is proportional to force F by a constant M:

$$F = M\delta \tag{1.4}$$

This represents a perfectly linear elastic material. The response to a force is instantaneous and the deformation is completely recovered when the force is removed. Thus, the Hookean element represents a perfectly linear elastic material.

A Newtonian element models a perfectly viscous material and is modeled as a dashpot or shock absorber as seen in Figure 1.10(b). The deformation for a given level of force is proportional to the amount of time the force is applied. Hence, the rate of deformation, for a constant force, is a constant β:

$$F = \beta \dot{\delta} \tag{1.5}$$

The dot above the δ defines this as the rate of deformation with respect to time. If $\delta=0$ at time $t=0$ when a constant force F is applied, the deformation at time t is

$$\delta = \frac{Ft}{\beta} \tag{1.6}$$

When the force is removed, the specimen retains the deformed shape. There is no recovery of any of the deformation.

The St. Venant element, as seen in Figure 1.10(c), has the characteristics of a sliding block that resists movement by friction. When the force F applied to the element is less than the critical force F_O, there is no movement. If the force is increased to overcome the static friction, the element will slide and continue to slide as long as the force is applied. This element is unrealistic, since any sustained force sufficient to cause movement would cause the block to accelerate. Hence, the St. Venant element is always used in combination with the other basic elements.

The basic elements are usually combined in parallel or series to model material response. Figure 1.11 shows the three primary two-component models: the Maxwell, Kelvin, and Prandtl models. The Maxwell and Kelvin models have a spring and dashpot in series and parallel, respectively. The Prandtl model uses a spring and St. Venant elements in series.

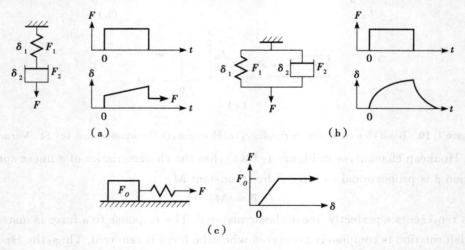

Figure 1.11 Two-element rheological models: (a) Maxwell, (b) Kelvin, and (c) Prandtl.

In the Maxwell model [Figure 1.11(a)], the total deformation is the sum of the deformations of the individual elements. The force in each of the elements must be equal to the total force ($F=F_1=F_2$). Thus, the equation for the total deformation at any time after a constant load is applied is simply

$$\delta = \delta_1 + \delta_2 = \frac{F}{M} + \frac{Ft}{\beta} \qquad (1.7)$$

In the Kelvin model [Figure 1.11(b)] the deformation of each of the elements must be equal at all times, due to the way the model is formulated. Thus, the total deformation is equal to the deformation of each element ($\delta=\delta_1=\delta_2$). Since the elements are in parallel, they will share the force such that the total force is equal to the sum of the force in each element. If $\delta=0$ at time $t=0$ when a constant force F is applied, Equation 1.4 then requires zero force in the spring. Hence, when the load is initially applied, before any deformation takes place, all of the force must be in the dashpot. Under constant force, the deformation of the dashpot must increase, since there is force on the element. However, this also requires deformation of the spring, indicating that some of the force is carried by the spring. In fact, with time, the amount of force in the dashpot decreases and the force in the spring increases. The proportion is fixed by the fact that the sum on the forces in the two elements must be equal to the total force. After a sufficient amount of time, all of the force will be transferred to the spring and the model will stop deforming. Thus, the maximum deformation of the Kelvin model is $\delta=F/M$. Mathematically, the equation for the deformation in a Kelvin model is derived as

$$F = F_1 + F_2 = M\delta + \beta\dot{\delta} \qquad (1.8)$$

Integrating Equation 1.8, using the limits that $\delta=0$ at $t=0$, and solving for the deformation δ at time t results in

$$\delta = \left(\frac{F}{M}\right)(1 - e^{-Mt/\beta}) \qquad (1.9)$$

The Prandtl model [Figure 1.11(c)] consists of St. Venant and Hookean bodies in series. The Prandtl model represents a material with an elastic-perfectly plastic response. If a small load is applied, the material responds elasticly until it reaches the yield point, after which the material exhibits plastic deformation.

Neither the Maxwell nor Kelvin model adequately describes the behavior of some common engineering materials, such as asphalt concrete. However, the Maxwell and the Kelvin models can be put together in series, producing the Burgers model, which can be used to describe simplistically the behavior of asphalt concrete. As shown in Figure 1.12, the Burgers model is generally drawn as a spring in series with a Kelvin model in series with a dashpot. The total deformation, when $\delta=0$ at time $t=0$, is then the sum of the deformations of these three elements:

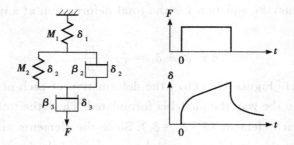

Figure 1.12 Burgers model of viscoelastic materials.

$$\delta = \delta_1 + \delta_2 + \delta_3 = \frac{F}{M_1} + \left(\frac{F}{M_2}\right)(1 - e^{-M_2 t/\beta_2}) + \frac{Ft}{\beta_3} \quad (1.10)$$

The deformation-time diagram for the loading part of the Burgers model demonstrates three distinct phases of behavior. The first phase is the instantaneous deformation of the spring when the load is applied. The second phase is the combined deformation of the Kelvin model and the dashpot. In the third phase, after the Kelvin model reaches maximum deformation, there is a continued deformation of the dashpot at a constant rate of deformation. The unloading part of the Burgers model follows similar behavior.

Some materials require more complicated rheological models to represent their response. In such cases, a number of Maxwell models can be combined in parallel to form the generalized Maxwell model, or a number of Kelvin models in series can be used to form the generalized Kelvin model.

The use of rheological models requires quantifying material parameters associated with each model. Laboratory tests, such as creep tests, can be used to obtain deformation-time curves from which material parameters can be determined.

Fig. SP1.4 the time-dependent response of materials, they can be used only to represent uniaxial responses. The three-dimensional behavior of materials and the Poisson's effect cannot be represented by these models.

Sample Problem 1.4

Derive the response relation for the model shown in Fig. SP1.4 assuming that the force F is constant and instantaneously applied.

Fig. SP1.4

Solution

$$\text{For } F \leqslant F_O : \delta = F/M$$
$$\text{For } F > F_O : \text{movement}$$

1.2.8 Temperature and Time Effects

The mechanical behavior of all materials is affected by temperature. Some materials, however, are more susceptible to temperature than others. For example, viscoelastic materials, such as plastics and asphalt, are greatly affected by temperature, even if the temperature is changed by only a few degrees. Other materials, such as metals or concrete, are less affected by temperatures, especially when they are near ambient temperature.

Ferrous metals, including steel, demonstrate a change from ductile to brittle behavior as the temperature drops below the transition temperature. This change from ductile to brittle behavior greatly reduces the toughness of the material. While this could be determined by evaluating the stress-strain diagram at different temperatures, it is more common to evaluate the toughness of a material with an impact test that measures the energy required to fracture a specimen. Figure 1.13 shows how the energy required to fracture a mild steel changes with temperature (Flinn and Trojan 1986). The test results seen in Figure 1.13 were achieved by applying impact forces on bar specimens with a "defect" (a simple V notch) machined into the specimens (ASTM E23). During World War II, many Liberty ships sank because the steel used in the ships met specifications at ambient temperature, but became brittle in the cold waters of the North Atlantic.

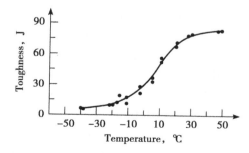

Figure 1.13 Fracture toughness of steel under impact testing.

In addition to temperature, some materials, such as viscoelastic materials, are affected by the load duration. The longer the load is applied, the larger is the amount of deformation or creep. In fact, increasing the load duration and increasing the temperature cause similar material responses. Therefore, temperature and time can be interchanged. This concept is very useful in running some tests. For example, a creep test on an asphalt concrete specimen can be performed with short load durations by increasing the temperature of the material. A time-temperature shift factor is then used to adjust the results for lower temperatures.

Viscoelastic materials are not only affected by the duration of the load, but also by the rate of load application. If the load is applied at a fast rate, the material is stiffer than if the load is applied at a slow rate. For example, if a heavy truck moves at a high speed on an asphalt pavement, no permanent deformation may be observed. However, if the same truck is parked on an asphalt pavement on a hot day, some permanent deformations on the pavement surface may be observed.

1.2.9 Failure and Safety

Failure occurs when a member or structure ceases to perform the function for which it was designed. Failure of a structure can take several modes, including fracture fatigue, general yielding, buckling, and excessive deformation. Fracture is a common failure mode. A brittle material typically fractures suddenly when the static stress reaches the strength of the material, where the strength is defined as the maximum stress the material can carry. On the other hand, a ductile material may fracture due to excessive plastic deformation.

Many structures, such as bridges, are subjected to repeated loadings, creating stresses that are less than the strength of the material. Repeated stresses can cause a material to fail or fatigue, at a stress well below the strength of the material. The number of applications a material can withstand depends on the stress level relative to the strength of the material. As shown in Figure 1.14, as the stress level decreases, the number of applications before failure increases. Ferrous metals have an apparent endurance limit, or stress level, below which fatigue does not occur. The endurance limit for steels is generally in the range of one-quarter to one-half the ultimate strength (Flinn and Trojan 1986). Another example of a structure that may fail due to fatigue is pavement. Although the stresses applied by traffic are typically much less than the strength of the material, repeated loadings may eventually lead to a loss of the structural integrity of the pavement surface layer, causing fatigue cracks as shown in Figure 1.15.

Another mode of failure is general yielding. This failure happens in ductile materials, and it spreads throughout the whole structure, which results in a total collapse.

Long and slender members subjected to axial compression may fail due to buckling. Although the member is intended to carry axial compressive loads, a small lateral force might be applied, which causes deflection and eventually might cause failure.

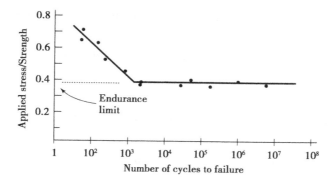

Figure 1.14 An example of endurance limit under repeated loading.

Figure 1.15 Fatigue failure of asphalt pavement due to repeated traffic loading.

Sometimes excessive deformation (elastic or plastic) could be defined as failure, depending on the function of the member. For example, excessive deflections of floors make people uncomfortable and, in an extreme case, may render the building unusable even though it is structurally sound.

To minimize the chance of failure, structures are designed to carry a load greater than the maximum anticipated load. The factor of safety (FS) is defined as the ratio of the stress at failure to the allowable stress for design (maximum anticipated stress):

$$FS = \frac{\sigma_{\text{failure}}}{\sigma_{\text{allowable}}} \qquad (1.11)$$

where σ_{failure} is the failure stress of the material and $\sigma_{\text{allowable}}$ is the allowable stress for design.

Typically, the larger the factor of safety, the larger is the required cross section of the structure and, consequently, the higher is the cost. The proper value of the factor of safety varies from one structure to another and depends on several factors, including the

- cost of unpredictable failure in lives, dollars, and time,
- variability in material properties,
- degree of accuracy in considering all possible loads applied to the structure, such as earthquakes,
- possible misuse of the structure, such as improperly hanging an object from a truss roof,
- degree of accuracy of considering the proper response of materials during design, such as assuming elastic response although the material might not be perfectly elastic.

1.3 Nonmechanical Properties

Nonmechanical properties refer to characteristics of the material, other than load response, that affect selection, use and performance. There are several types of properties that are of interest to engineers, but those which are of the greatest concern to civil engineers are density, thermal properties, and surface characteristics.

1.3.1 Density and Unit Weight

In many structures, the dead weight of the materials in the structure significantly contributes to the total design stress. If the weight of the materials can be reduced, the size of the structural members can be also reduced. Thus, the weight of the materials is an important design consideration. In addition, in the design of asphalt and concrete mixes, the weight-volume relationship of the aggregates and binders must be used to select the mix proportions.

There are three general terms used to describe the mass, weight, and volume relationship of materials. Density is the mass per unit volume of material. Unit weight is the weight per unit volume of material. By manipulation of units, it can be shown that

$$\gamma = \rho g \qquad (1.12)$$

where
γ = unit weight
ρ = density
g = acceleration of gravity

Specific gravity is the ratio of the mass of a substance relative to the mass of an equal volume of water at a specified temperature. The density of water is 1 Mg/m³ in SI units and 62.4 lb/ft³ in English units at 4 ℃ (39.2 ℉). According to the definition, specific gravity is equivalent to the density of a material divided by the density of water. Since the density of water in the metric system has a numerical value of 1, the numerical value of density and specific gravity are equal. This fact is often used in the literature where density and specific gravity terms are used interchangeably.

For solid materials, such as metals, the unit weight, density, and specific gravity have definite numerical values. For other materials such as wood and aggregates, voids in the materials require definitions for a variety of densities and specific gravities. As shown in Figure 1.16(a) and (b), the bulk volume aggregates will occupy depends on the compaction state of the material. In addition, the density of the material will change depending on how the volume of individual particles is measured. Several types of particle volume can be used, such as the total volume enclosed within the boundaries of the individual particles, volume not accessible to water or asphalt, and volume of solids, as seen in Figure 1.16(c), (d), and (e), respectively. These are important factors in the mix designs of portland cement concrete and asphalt concrete.

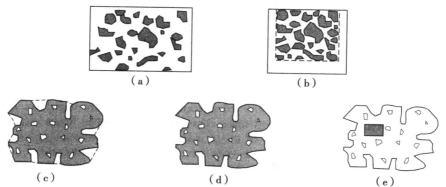

Figure 1.16 Definitions of volume used for determining density: (a) loose, (b) compacted, (c) total particle volume, (d) volume not accessible to water, and (e) volume of solids.

1.3.2 Thermal Expansion

Practically all materials expand as temperature increases and contract as temperature falls. The amount of expansion per unit length due to one unit of temperature increase is a material constant and is expressed as the coefficient of thermal expansion

$$\alpha_L = \frac{\delta L/\delta T}{L} \quad (1.13)$$

$$\alpha_V = \frac{\delta V/\delta T}{V} \qquad (1.14)$$

where

α_L = linear coefficient of thermal expansion
α_V = volumetric coefficient of thermal expansion
δL = change in the length of the specimen
δT = change in temperature
L = original length of the specimen
δV = change in the volume of the specimen
V = original volume of the specimen

For isotropic materials, $\alpha_V = 3\alpha_L$.

The coefficient of thermal expansion is very important in the design of structures. Generally, structures are composed of many materials that are bound together. If the coefficients of thermal expansion are different, the materials will strain at different rates. The material with the lesser expansion will restrict the straining of other materials. This constraining effect will cause stresses in the materials that can lead directly to fracture.

Stresses can also be developed as a result of a thermal gradient in the structure. As the temperature outside the structure changes and the temperature inside remains constant, a thermal gradient develops. When the structure is restrained from straining, stress develops in the material. This mechanism has caused brick facades on buildings to fracture and, in some cases, fall off the structure. Also, since concrete pavements are restrained from movement, they may crack in the winter due to a drop in temperature and may "blow up" in the summer due to an increase in temperature. Joints are, therefore, used in buildings, bridges, concrete pavements, and various structures to accommodate this thermal effect.

Sample Problem 1.5

A steel bar with a length of 3 m, diameter of 25 mm, modulus of elasticity of 207 GPa, and linear coefficient of thermal expansion of 0.000009 m/m/℃ is fixed at both ends when the ambient temperature is 40 ℃. If the ambient temperature is decreased to 15 ℃, what internal stress will develop due to this temperature change? Is this stress tension or compression? Why?

Solution

If the bar was fixed at one end and free at the other end, the bar would have contracted and no stresses would have developed. In that case, the change in length can be calculated by using Equation 1.13 as follows:

$$\delta L = \alpha_L \times \delta T \times L = 0.000009 \times (-25) \times 3 = -0.000675 \text{ m}$$
$$\varepsilon = \delta L / L = -0.000675 / 3 = -0.000225 \text{ m/m}$$

Since the bar is fixed at both ends, the length of the bar will not change. Therefore, a tensile stress will develop in the bar as follows:

$$\sigma = \varepsilon E = 0.000225 \times 207\,000 = 46.575 \text{ MPa}$$

The stress will be tension; in effect, the length of the bar at 15 ℃ without restraint would be 2.999325 m and the stress would be zero. Restraining the bar into a longer condition requires a tensile force.

1.3.3 Surface Characteristics

The surface properties of materials of interest to civil engineers include corrosion and degradation, the ability of the material to resist abrasion and wear, and surface texture.

Corrosion and Degradation Nearly all materials deteriorate over their service lives. The mechanisms contributing to the deterioration of a material differ depending on the characteristics of the material and the environment. Crystalline materials, such as metals and ceramics, deteriorate through a corrosion process in which there is a loss of material, either by dissolution or by the formation of nonmetallic scale or film. Polymers, such as asphalt, deteriorate by degradation of the material, including the effects of solvent and ultraviolet radiation on the material.

The protection of materials from environmental degradation is an important design concern, especially when the implications of deterioration and degradation on the life and maintenance costs of the structure are considered. The selection of a material should consider both how the material will react with the environment and the cost of preventing the resulting degradation.

Abrasion and Wear Resistance Since most structures in civil engineering are static, the abrasion or wear resistance is of less importance than in other fields of engineering. For example, mechanical engineers must be concerned with the wear of parts in the design of machinery. This is not to say that wear resistance can be totally ignored in civil engineering. Pavements must be designed to resist the wear and polishing from vehicle tires in order to provide adequate skid resistance for braking and turning. Resistance to abrasion and wear is, therefore, an important property of aggregates used in pavements.

Surface Texture The surface texture of some materials and structures is of importance to civil engineers. For example, smooth texture of aggregate particles is needed in portland cement concrete to improve workability during mixing and placing. In contrast, rough texture of aggregate particles is needed in asphalt concrete mixtures to provide a stable pavement layer that resists deformation under the action of load. Also, a certain level of surface texture is needed in the pavement surface to provide adequate friction resistance and prevent skidding of vehicles when the pavement is wet.

1.4 Production and Construction

Even if a material is well suited to a specific application, production and construction considerations may block the selection of the material. Production considerations include the availability of the material and the ability to fabricate the material into the desired shapes and required specifications. Construction considerations address all the factors that relate to the ability to fabricate and erect the structure on site. One of the primary factors is the availability of a trained work force. For example, in some cities high-strength concrete is used for skyscrapers, whereas in other cities steel is the material of choice. Clearly, either concrete or steel can be used for high-rise buildings. Regional preferences for one material develop as engineers in the region become comfortable and confident in designing with one of the materials and constructors respond with a trained work force and specialized equipment.

1.5 Aesthetic Characteristics

The aesthetic characteristics of a material refer to the appearance of the material. Generally, this characteristic is the responsibility of the architect. However, the civil engineer is responsible for working with the architect to ensure that the aesthetic characteristics of the facility are compatible with the structural requirements. During the construction of many public projects, a certain percentage of the capital budget typically goes toward artistic input. The collaboration between the civil engineer and the architect is greatly encouraged, and the result can increase the value of the structure.

In many cases, the mix of artistic and technical design skills makes the project acceptable to the community. In fact, political views are often more difficult to deal with than technical design problems. Thus, engineers should understand that there are many factors beyond the technical needs that must be considered when selecting materials and designing public projects.

1.6 Material Variability

It is essential to understand that engineering materials are inherently variable. For example, steel properties vary depending on chemical composition and method of manufacture. Concrete properties change depending on type and amount of cement, type of aggregate, air content, slump, method of curing, etc. The properties of asphalt concrete vary depending on the binder amount and type, aggregate properties and gradation, amount of compaction, and age. Wood properties vary depending on the tree species, method of cut, and moisture content. Some materials are more homogeneous than others, depending on the nature of the material and the method of manufacturing. For example, the variability of the yield strength of one type of steel is less than the variability of the compressive strength of one batch of concrete. Therefore, variability is an important parameter in defining the quality of civil engineering materials.

When materials from a particular lot are tested, the observed variability is the cumulative effect of three types of variance: the inherent variability of the material, variance caused by the sampling method, and variance associated with the way the tests are conducted. Just as materials have an inherent variability, sampling procedures and test methods can produce variable results. Frequently, statisticians call variance associated with sampling and testing error. However, this does not imply the sampling or testing was performed incorrectly. When an incorrect procedure is identified, it is called a blunder. The goal of a sampling and testing program is to minimize sampling and testing variance so the true statistical features of the material can be identified.

The concepts of precision and accuracy are fundamental to the understanding of variability. Precision refers to the variability of repeat measurements under carefully controlled conditions. Accuracy is the conformity of results to the true value or the absence of bias. Bias is a tendency of an estimate to deviate in one direction from the true value. In other words, bias is a systematic error between a test value and the true value. A simple analogy to the relationship between precision and accuracy is the target shown in Figure 1.17. When all shots are concentrated at one location away from the center, that indicates good precision and poor accuracy (biased) [Figure 1.17(a)]. When shots are scattered around the center, that indicates poor precision and good accuracy [Figure 1.17(b)]. Finally, good precision and good accuracy are obtained if all shots are concentrated close to the center [Figure 1.17(c)](Burati and Hughes 1990). Many standardized test methods, such as those of the American Society for Testing and Materials (ASTM) and the American Association of State Highway and Transportation Officials (AASHTO), contain precision and bias statements. These statements provide the limits of acceptable test results variability. Laboratories are usually required to demonstrate testing competence and can be certified by the American Material Reference Laboratory (AMRL).

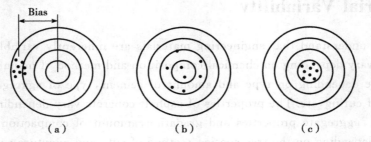

Figure 1.17 Exactness of measurements: (a) precise but not accurate, (b) accurate but not precise, and (c) precise and accurate.

1.6.1 Sampling

Typically, samples are taken from a lot or population, since it is not practical or possible to test the entire lot. By testing sufficient samples, it is possible to estimate the properties of the entire lot. In order for the samples to be valid they must be randomly selected. Random sampling requires that all elements of the population have an equal chance for selection. Another important concept in sampling is that the sample must be representative of the entire lot. For example, when sampling a stockpile of aggregate, it is important to collect samples from the top, middle, and bottom of the pile and to combine them, since different locations within the pile are likely to have different aggregate sizes. The sample size needed to quantify the characteristics of a population depends on the variability of the material properties and the confidence level required in the evaluation.

Statistical parameters describe the material properties. The mean and the standard deviation are two commonly used statistics. The arithmetic mean is simply the average of test results of all specimens tested. It is a measure of the central tendency of the population. The standard deviation is a measure of the dispersion or spread of the results. The equations for the mean $\overline{x}$ and standard deviation s of a sample are

$$\overline{x} = \frac{\sum_{i=1}^{n} x_i}{n} \qquad (1.15)$$

$$s = \left[\frac{\sum_{i=1}^{n}(x_i - \overline{x})^2}{n-1} \right]^{1/2} \qquad (1.16)$$

where n is the sample size. The mean and standard deviation of random samples are estimates of the mean and standard deviation, respectively, of the population.

1.6.2 Normal Distribution

The normal distribution is a symmetrical function around the mean, as shown in Figure 1.18. The normal distribution describes many populations that occur in nature, research, and industry, including material properties. The area under the curve between any two values represents the probability of occurrence of an event of interest. Expressing the results in terms of mean and standard deviation, it is possible to determine the probabilities of an occurrence of an event. For example, the probability of occurrence of an event between the mean and ±1 standard deviation is 68.3%, between the mean and ±2 standard deviations is 95.5%, and between the mean and ±3 standard deviations is 99.7%. If a materials engineer tests 20 specimens of concrete and determines the average as 22 MPa and the standard deviation as 3 MPa, the statistics will show that 95.5% of the time the true mean of the population will be in the range of 22 ±(2×3), or 16 to 28 MPa.

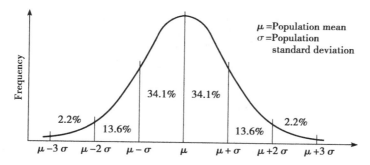

Figure 1.18 A normal distribution.

1.6.3 Control Charts

Control charts have been used in manufacturing industry and construction applications to verify that a process is in control. It is important to note that control charts do not get or keep a process under control; they provide only a visual warning mechanism to identify when a contractor or material supplier should look for possible problems with the process. Control charts have many benefits (Burati and Hughes 1990), such as

◆ detect trouble early

- decrease variability
- establish process capability
- reduce price adjustment cost
- decrease inspection frequency
- provide a basis for altering specification limits
- provide a permanent record of quality
- provide basis for acceptance
- instill quality awareness

There are many types of control charts, the simplest of which plots individual results in chronological order. For example, Figure 1.19 shows a control chart of the compressive strength of concrete specimens tested at a ready-mix plant. The control chart can also show the specification tolerance limits so that the operator can identify when the test results are out of the specification requirements. Although this type of control chart is useful, it is based on a sample size of one, and it therefore fails to consider variability within the sample.

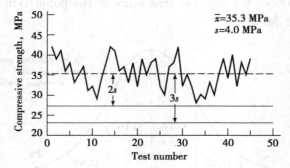

Figure 1.19 Control chart of compressive strength of concrete specimens.

Statistical control charts can be developed, such as the control chart for means (X-bar chart) and the control chart for the ranges (R chart) in which the means or the ranges of the test results are chronologically plotted. Figure 1.20(a) shows a control chart for the moving average of each three consecutive compressive strength tests. For example, the first point represents the mean of the first three tests, the second point represents the mean of tests two through four, and so on. Figure 1.20(b) shows a control chart for the moving range of each three consecutive compressive strength tests. The key element in the use of statistical control charts is the proper designation of the control limits that are set for a given process. These control limits are not necessarily the same as the tolerance or specification limits and can be set using probability functions. For example, the control chart for means relies on the fact that, for a normal distribution, essentially all of the values fall within ±3 standard deviations from the mean. Thus, control limits can be set between ±3 standard deviations from the mean. Warning limits to identify potential problems are sometimes set at ±2 standard deviations from the mean.

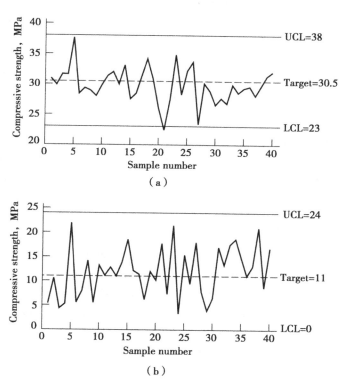

Figure 1.20 Statistical control charts: (a) X-bar chart and (b) R chart. UCL indicates upper control limit; LCL indicates lower control limit.

Observing the trend of means or ranges in statistical control charts can help eliminate production problems and reduce variability. Figure 1.21 shows possible trends of means and ranges in statistical control charts (Burati and Hughes 1990). Figure 1.21(a) shows sustained sudden shift in the mean. This could indicate a change of a material supplier during the project. A gradual change in the mean, as illustrated in Figure 1.21(b), could indicate a progressive change brought on by machine wear. An irregular shift in the mean, as shown in Figure 1.21(c), may indicate that the operator is making continuous, but unnecessary, adjustments to the process settings. Figure 1.21(d) shows a sudden change in range, which could also indicate a change of a material supplier during the project. Figure 1.21(e) shows a gradual increase in the range, which may indicate machine wear. Finally, Figure 1.21(f) shows an irregular shift in both mean and range, which indicates a flawed process.

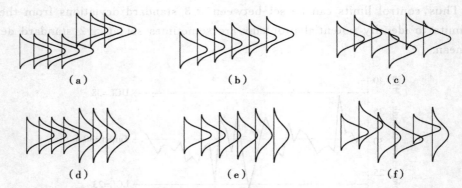

Figure 1.21 Possible trends of means and ranges in statistical control charts: (a) sudden change in mean, (b) gradual change in mean, (c) irregular change in mean, (d) sudden change in range, (e) gradual change in range, and irregular change in mean and range.

1.6.4 Experimental Error

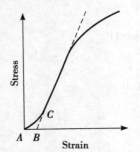

Figure 1.22 Correction of toe region in stress-strain curve.

When specimens are tested in the laboratory, inaccuracy could occur due to machine or human errors. For example, Figure 1.22 shows a stress-strain curve in which a toe region (AC) that does not represent a property of the material exists. This toe region is an artifact caused by taking up slack and alignment or seating of the specimen. In order to obtain correct values of such parameters as modulus, strain, and offset yield point, this artifact must be compensated for in order to give the corrected zero point on the strain axis. This is accomplished by extending the linear portion of the curve backward until it meets the strain axis at point B. In this case, point B is the corrected zero strain point from which all strains must be measured. In the case of a material that does not exhibit any linear region, a similar correction can be made by constructing a tangent to the maximum slope at the inflection point and extending it until it meets the strain axis.

1.7 Laboratory Measuring Devices

Laboratory tests measure material properties. Frequently, specimens are made of the material in question and tested in the laboratory to measure their response to the applied forces or to certain environmental conditions. These tests require the measurement of certain parameters such as time, deformation, or force. Some of these parameters are measured directly, while others are measured indirectly by relating parameters to each other. Length and deformation can be measured directly using simple devices such as rulers, dial gauges, and calipers. In other cases, indirect measurements are made by measuring electric voltage and relating it to deformation, force, stress, or strain. Examples of such devices include linear variable differential transformers (LVDTs), strain gauges, and load cells. Noncontact deformation measuring devices using lasers and various optical devices are also available. Electronic measuring devices can easily be connected to chart recorders, digital readout devices, or computers, where the results can be easily displayed and processed.

Each device has a certain *sensitivity*, which is the smallest value that can be read on the device's scale. Sensitivity should not be mistaken for accuracy or precision. Magnification can be designed into a gauge to increase its sensitivity, but wear, friction, noise, drift, and other factors may introduce errors that limit the accuracy and precision.

Measurement accuracy cannot exceed the sensitivity of the measuring device. For example, if a stopwatch with a sensitivity of 0.01 second is used to measure time, the smallest time interval that can be recorded is also 0.01 second. The selection of the measuring device and its sensitivity depends on the required accuracy of measurement. The required accuracy, on the other hand, depends on the significance and use of the measurement. For example, when expressing distance of travel from one city to another, an accuracy of 1 kilometer or even 10 kilometers may be meaningful. In contrast, manufacturing a computer microchip may require an accuracy of one-millionth of a meter or better. In engineering tests, the accuracy of measurement must be determined in advance to ensure proper use of such measurements and, at the same time, to avoid unnecessary effort and expense during testing. Many standardized test methods, such as those of ASTM and AASHTO, state the sensitivity of the measuring devices used in a given experiment. In any case, care must be taken to ensure proper calibration, connections, use, and interpretation of the test results of various measuring devices.

Next we will briefly describe measuring devices commonly used in material testing, such as dial gauges, linear variable differential transformers (LVDTs), strain gauges, proving rings, and load cells.

1.7.1 Dial Gauge

Dial gauges are used in many laboratory tests to measure deformation. The dial gauge is attached at two points, between which the relative movement is measured. Most of the dial gauges include two scales with two different pointers, as depicted in Figure 1.23. The smallest division of the large scale determines the sensitivity of the device and is usually recorded on the face of the gauge. One division of the small pointer corresponds to one full rotation of the large pointer. The full range of the small pointer determines the range of measurement of the dial gauge. Dial gauges used in civil engineering material testing frequently have sensitivities ranging from 0.1 mm to 0.002 mm. The dial gauge shown in Figure 1.23 has a sensitivity of 0.001 in. and a range of 1 in. The gauge can be "zeroed" by rotating the large scale in order to start the reading at the current pointer position.

Figure 1.23 Dial gauge.

Figure 1.24 Extensometer with a dial gauge.

Dial gauges can be attached to frames or holders with different configurations to measure the deformation of a certain gauge length or the relative movement between two points. For example, the *extensometer* shown in Figure 1.24 is used to measure the deformation of the gauge length of a metal bar during the tensile test. Note that because of the extensometer configuration shown in the figure, the deformation of the bar is one-half of the reading indicated by the dial gauge.

1.7.2 Linear Variable Differential Transformer (LVDT)

The linear variable differential transformer (transducer), or LVDT, is an electronic device commonly used in laboratory experiments to measure small movements or deformations of specimens. The LVDT consists of a nonmagnetic shell and a magnetic core. The shell contains one primary and two secondary electric coils, as illustrated in Figure 1.25. An electric voltage is input to the LVDT and an output voltage is obtained. When the core is in the null position at the center of the shell, the output voltage is zero. When the core is moved slightly in one direction, an output voltage is obtained. Displacing the core in the opposite direction produces an output voltage with the opposite sign. The relationship between the core position and the output voltage is linear within a certain range determined by the manufacturer. If this relation is known, the displacement can be determined by measuring the output voltage using a voltmeter or a readout device. LVDTs can measure both static and dynamic movements.

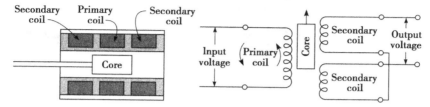

Figure 1.25 LVDT circuit.

LVDTs vary widely in sensitivity and range. The sensitivity of commercially available LVDTs ranges from 0.003 to 0.25 V/mm (0.08 to 6.3 V/in.) of displacement per volt of excitation. Normal excitation supplied to the primary coil is 3 Vac with a frequency ranging from 50 Hz to 10 kHz. If 3 V is used, the most sensitive LVDTs provide an output of 18.9 mV/mm (Dally and Riley 1991). In general, very sensitive LVDTs have small linear ranges, whereas LVDTs with greater linear ranges are less sensitive. The sensitivity and the linear range needed depend on the accuracy and the amount of displacement required for the measurement.

Before use, the LVDT must be calibrated to determine the relation between the output voltage and the displacement. A calibration device consisting of a micrometer, a voltmeter, and a holder is used to calibrate the LVDT, as shown in Figure 1.26.

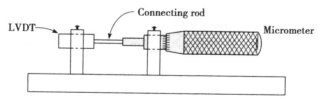

Figure 1.26 LVDT calibration device.

The shell and the core of the LVDT can be either separate or attached in a spring-loaded arrangement (Figure 1.27). When the former type is used, a nonmagnetic threaded connecting rod is attached to the core used to attach the LVDT to the measured object. In either case, to measure the relative movement between two points, the core is attached to one point and the shell is attached to the other point. When the distance between the two points changes, the core position changes relative to the shell, proportionally altering the output voltage. Figure 1.28 shows an extensometer with an LVDT that can be used to measure the deformation of a metal rod during the tensile test.

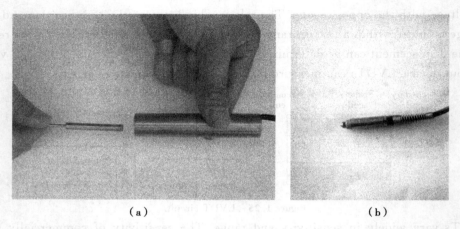

Figure 1.27 Types of LVDT: (a) non-spring loaded (core and connecting rod taken out of the shell), and (b) spring loaded (core and connecting rod inside the shell).

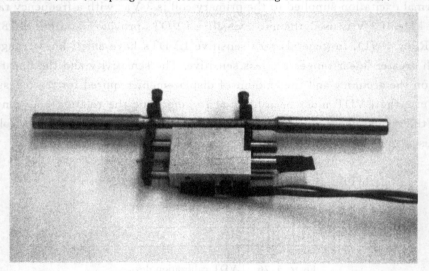

Figure 1.28 Extensometer with an LVDT.

1.7.3 Strain Gauge

Strain gauges are used to measure small deformations within a certain gauge length. There are several types of strain gauge, but the most dominant type is the electrical strain gauge, which consists of a foil or wire bonded to a thin base of plastic or paper (Figure 1.29). An electric current is passed through the element (foil or wire). As the element is strained, its electrical resistance changes proportionally. The strain gauge is bonded via an adhesive to the surface on which the strain measurement is desired. As the surface deforms, the strain gauge also deforms and, consequently, the resistance changes. Since the amount of resistance change is very small, an ordinary ohmmeter cannot be used. Therefore, special electric circuits, such as the Wheatstone bridge, are used to detect the change in resistance (Dally and Riley 1991).

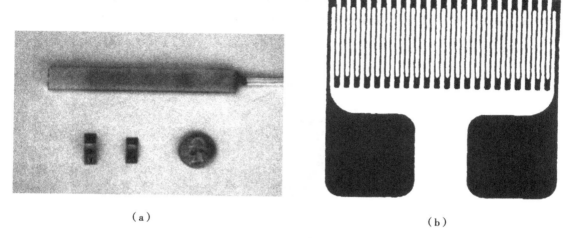

(a)   (b)

Figure 1.29 Strain gauges: (a) strain gauges with different sizes and (b) typical foil strain gauge.

Strain gauges are manufactured with different sizes, but the most convenient strain gauges have a gauge length of about 5 mm to 15 mm (1/4 in. to 1/2 in.). Larger strain gauges can also be made and used in some applications.

A wire gauge consists of a length of very fine wire (about 0.025 mm diameter) that is looped into a pattern. A foil gauge is made by etching a pattern on a very thin metal foil (about 0.0025 mm thick). Foils or wires are made in a great variety of shapes, sizes, and types and are bonded to a plastic or paper base. When the strain gauge is bonded to the object, it is cemented firmly with the foil or wire side out. Foil-plastic gauges are more commonly used than wire gauges.

When using strain gauges, it is very important to have a tight bond between the gauge and the member. The surface must be carefully cleaned and prepared and the adhesive must be properly applied and cured. The adhesive must be compatible with the material being tested.

1.7.4 Proving Ring

Proving rings are used to measure forces in many laboratory tests. The proving ring consists of a steel ring with a dial gauge attached, as shown in Figure 1.30. When a force is applied on the proving ring, the ring deforms, as measured with the dial gauge. If the relation between the force and dial gauge reading is known, the proving ring can be used to measure the applied force. Therefore, the proving ring comes with a calibration relationship, either in a form of a linear equation or a table, that allows the user to determine the magnitude of the force based on measuring the deformation of the proving ring. To avoid damage, it is important not to apply a force on the proving ring higher than the capacity specified by the manufacturer. Moreover, periodic calibration of the proving ring is advisable to insure proper measurements.

Figure 1.30 Proving ring.

1.7.5 Load Cell

The load cell is an electronic force-measuring device used for many laboratory tests. In this device, strain gauges are attached to a member within the load cell, which is subjected to either axial loading or bending. An electric voltage is input to the load cell and an output voltage is obtained. If the relation between the force and the output voltage is known, the force can easily be determined by measuring the output voltage.

Load cells are manufactured in different shapes and load capacities. Figure 1.31(a) illustrates a tensile load cell fabricated by mounting four strain gauges on the central region of a tension specimen. Figure 1.31(b) shows an S-shaped load cell in which strain gauges are bonded to the central portion and calibrated to measure the force applied on the top and bottom of the load cell. Figure 1.31(c) illustrates a diaphragm strain gauge bonded to the inside surface of an enclosure that measures the amount of pressure applied on the load cell.

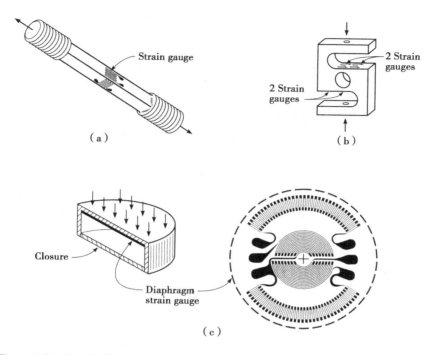

Figure 1.31 Load cells: (a) strain gauges on a tension rod, (b) strain gauges on an S-shaped element, and (c) diaphragm strain gauge on a closure.

Load cells must be regularly calibrated using either dead loads or a calibrated loading machine. Care must be taken not to overload the load cell. If the load applied on the load cell exceeds the capacity recommended by the manufacturer, permanent deformation can develop, ruining the load cell.

SUMMARY

Civil and construction engineers are involved in the selection of construction materials with the mechanical properties needed for each project. In addition, the selection process must weigh other factors beyond the material's ability to carry loads. Economics, production, construction, maintenance, and aesthetics must all be considered when selecting a material.

Lately, in all fields of engineering, there has been tremendous growth in the use of new high-performance materials. For example, in the automotive industry, applications of ceramics and plastics are increasing as manufacturers strive for better performance, economy, and safety, while pushing to reduce emissions. Likewise, civil and construction engineers are continuously looking for materials with better quality and higher performance. Advanced composite materials, geotextiles, and various synthetic products are currently competing with traditional civil engineering materials. Although traditional materials such as steel, concrete, wood, and asphalt will continue to be used for some time, improvements of these materials will proceed by changing the molecular structure of such materials and using modifiers to improve their performance. Examples of such improvements include fiber-reinforced concrete, polymer-modified concrete and asphalt, low temperature-susceptible asphalt binder, high-early-strength concrete, superplasticizers, epoxy-coated steel reinforcement, synthetic bar reinforcement, rapid-set concrete patching compounds, prefabricated drainage geocomposites, lightweight aggregates, fire-resistant building materials, and earthquake-resistant joints. Civil engineers are also recycling old materials in an effort to save materials cost, reduce energy, and improve the environment.

QUESTIONS AND PROBLEMS

1.1 A material has the stress-strain behavior shown in Figure P1.1. What is the material strength at rupture? What is the toughness of this material?

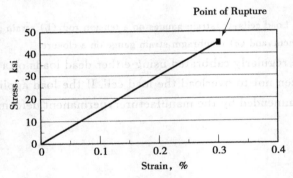

Figure P1.1

1.2 A cylindrical rod with a length of 380 mm and a diameter of 10 mm is to be subjected to a tensile load. The rod must not experience plastic deformation or an increase in length of more than 0.9 mm when a load of 24.5 kN is applied. Which of the four materials listed in the accompanying table are possible candidates? Justify your answer.

Material	Elastic Modulus, GPa	Yield Strength, MPa	Tensile Strength, MPa
Copper	110	248	289
Aluminum alloy	70	255	420
Steel	207	448	551
Brass alloy	101	345	420

1.3 The stress-strain relation shown in Figure P1.3 was obtained during the tensile test of an aluminum alloy specimen.

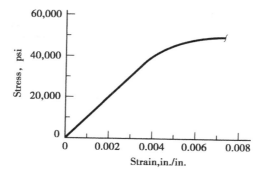

Figure P1.3

Determine the following:
a. Young's modulus within the linear portion
b. Tangent modulus at a stress of 45,000 psi
c. Yield stress using an offset of 0.002 strain
d. If the yield stress in part c is considered failure stress, what is the maximum working stress to be applied to this material if a factor of safety of 1.5 is used?

1.4 An elastoplastic material with strain hardening has the stress-strain relation shown in Figure P1.4. The yield point corresponds to 600 MPa stress and 0.003 m/m strain.

a. If a bar made of this material is subjected to a stress of 650 MPa and then released, what is the permanent strain?

b. What is the percent increase in yield strength that is gained by the strain hardening shown in part (a)?

c. What is the percent increase in strength that is gained by the strain hardening shown in part (a)?

d. After strain hardening, if the material is subjected to a stress of 625 MPa, how much strain is obtained? Is this strain elastic, permanent, or a combination of both?

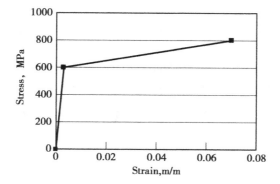

Figure P1.4

1.5 Derive the response relation for each of the models shown in Figure P1.5, assuming that the force F is constant and instantaneously applied.

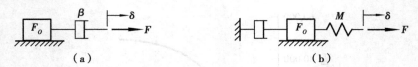

Figure P1.5

1.6 Briefly discuss the concept behind each of the following measuring devices:
   a. LVDT
   b. strain gauge
   c. proving ring
   d. load cell

References

Ashby, M. F. and D. R. H. Jones. *Engineering Materials: An Introduction to Their Properties and Applications*. New York: Pergamon Press, 1980.

Burati, J. L. and C. S. Hughes. *Highway Materials Engineering, Module I: Materials Control and Acceptance—Quality Assurance*. Publication No. FHWA-HI-90-004. Washington, DC: Federal Highway Administration, 1990.

Callister, W. D., Jr. *Materials Science and Engineering-An Introduction*. 6th ed. New York: John Wiley and Sons, 2003.

Dally. J. W. and W. F. Riley. *Experimental Stress Analysis*. 3d ed. New York: McGraw-Hill, 1991.

Flinn, R. A. and P. K. Trojan. *Engineering Materials and Their Applications*. 3d ed. Boston, MA: Houghton Mifflin, 1986.

Mehta, P. K. and P. J. M. Monteiro. *Concrete Structure, Properties, and Materials*. 2d ed. Upper Saddle River, NJ: Prentice Hall, 1993.

Polowski, N. H. and E. J. Ripling. *Strength and Structure of Engineering Materials*. Upper Saddle River, NJ: Prentice Hall, 1966.

Popov, E. P. *Introduction to Mechanics of Solids*. Upper Saddle River, NJ: Prentice Hall, 1968.

Richart, F. E., Jr., J. R. Hall, Jr., and R. D. Woods. *Vibrations of Soils and Foundations*. Upper Saddle River, NJ: Prentice Hall, 1970.

CHAPTER 2
STEEL

The use of iron dates back to about 1500 B. C. when primitive furnaces were used to heat the ore in a charcoal fire. Ferrous metals were produced on a relatively small scale until the blast furnace was developed in the 18th century. Iron products were widely used in the latter half of the 18th century and the early part of the 19th century. Steel production started in mid-1800s when the Bessemer converter was invented. In the second half of the 19th century, steel technology advanced rapidly due to the development of the basic oxygen furnace and continuous casting methods. More recently, computer-controlled manufacturing has increased the efficiency and reduced the cost of steel production.

Currently, steel and steel alloys are used widely in civil engineering applications. In addition, wrought iron is still used on a smaller scale for pipes, as well as for general blacksmith work. Cast iron is used for pipes, hardware, and machine parts not subjected to tensile or dynamic loading.

Steel products used in construction can be classified as follows:

1. structural steel for use in plates, bars, pipes, structural shapes, etc.
2. fastening products used for structural connections, including bolts, nuts and washers
3. reinforcing steel (rebars) for use in concrete reinforcement
4. miscellaneous products for use in such applications as forms and pans

Civil and construction engineers rarely have the opportunity to formulate steel with specific properties. Rather, they must select existing products from suppliers. Even the shapes for structural elements are generally restricted to those readily available from manufacturers. While specific shapes can be made to order, the cost to fabricate low-volume members is generally prohibitive. Therefore, the majority of civil engineering projects are designed using standard steel types and structural shapes.

Even though civil and construction engineers are not responsible for formulating steel products, they still must understand how steel is manufactured and treated and how it responds to loads and environmental conditions. This chapter reviews steel production, the iron-carbon phase diagram, heat treatment, steel alloys, structural steel, steel fasteners, and reinforcing steel. The chapter also presents common tests used to characterize the mechanical properties of steel. The topics of welding and corrosion of steel are also introduced.

2.1 Steel Production

The overall process of steel production is shown in Figure 2.1. This process consists of the following three phases:

1. reducing iron ore to pig iron
2. refining pig iron to steel
3. forming the steel into products

The materials used to produce pig iron are coal, limestone, and iron ore. The coal, after transformation to coke, supplies carbon used to reduce iron oxides in the ore. Limestone is used to help remove impurities. Prior to reduction, the concentration of iron in the ore is increased by crushing and soaking the ore. The iron is magnetically extracted from the waste, and the extracted material is formed into pellets and fired. The processed ore contains about 65% iron.

Reduction of the ore to pig iron is accomplished in a blast furnace. The ore is heated in the presence of carbon. Oxygen in the ore reacts with carbon to form gases. A flux is used to help remove impurities. The molten iron, with an excess of carbon in solution, collects at the bottom of the furnace. The impurities, slag, float on top of the molten pig iron.

The excess carbon, along with other impurities, must be removed to produce high-quality steel. Using the same refining process, scrap steel can be recycled. Three types of furnaces are used for refining pig iron to steel:

1. open hearth
2. basic oxygen
3. electric arc

The open hearth and basic oxygen furnaces remove excess carbon by reacting the carbon with oxygen to form gases. Lances circulate oxygen through the molten material. The process

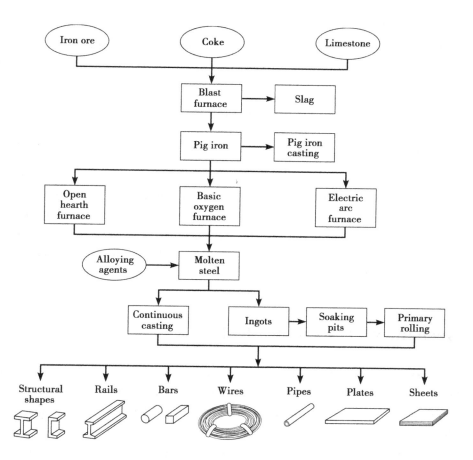

Figure 2.1 Conversion of raw material into different steel shapes.

is continued until all impurities are removed and the desired carbon content is achieved. Open hearth furnaces have been used since the early 1900s. Now, due to greater efficiency and productivity, basic oxygen furnaces are the industry standard for high-production mills. A basic oxygen furnace can refine 280,000 kg of steel in 25 minutes, compared with the eight hours it takes to refine the same quantity of steel in an open hearth furnace.

Electric furnaces use an electric arc between carbon electrodes to melt and refine the steel. These plants require a tremendous amount of energy and are primarily used to recycle scrap steel. Electric furnaces are frequently used in minimills, which produce a limited range of products. In this process, molten steel is transferred to the ladle. Alloying elements and additional agents can be added either in the furnace or the ladle.

During the steel production process, oxygen may become dissolved in the liquid metal. As the steel solidifies, the oxygen can combine with carbon to form carbon monoxide bubbles that are trapped in the steel and can act as initiation points for failure. Deoxidizing agents, such as aluminum, ferrosilicon and manganese, can eliminate the formation of the carbon monoxide bubbles. Completely deoxidized steels are known as *killed steels*. Steels that are generally killed include

- Those with a carbon content greater than 0.25%
- All forging grades of steels
- Structural steels with carbon content between 0.15 and 0.25 percent
- Some special steel in the lower carbon ranges

Regardless of the refining process, the molten steel, with the desired chemical composition, is then either cast into ingots (large blocks of steel) or cast continuously into a desired shape. Continuous casting is becoming the standard production method, since it is more energy efficient than casting ingots, as the ingots must be reheated prior to shaping the steel into the final product.

2.2 Mechanical Testing of Steel

Many tests are available to evaluate the mechanical properties of steel. This section summarizes some laboratory tests commonly used to determine properties required in product specifications. Test specimens can take several shapes, such as bar, tube, wire, flat section, and notched bar, depending on the test purpose and the application.

Certain methods of fabrication, such as bending, forming, and welding, or operations involving heating, may affect the properties of the material being tested. Therefore, the product specifications cover the stage of manufacture at which mechanical testing is performed. The properties shown by testing before the material is fabricated may not necessarily be representative of the product after it has been completely fabricated. In addition, flaws in the specimen or improper machining or preparation of the test specimen will give erroneous results (ASTM A370).

2.2.1 Tension Test

The tension test (ASTM E8) on steel is performed to determine the yield strength, yield point, ultimate (tensile) strength, elongation, and reduction of area. Typically, the test is performed at temperatures between 10 °C and 35 °C.

Table 2.1 Standard-Size Reinforcing Bars According to ASTM A615*

Bar Designation Number**	Nominal Mass (kg/m)	Nominal Dimensions***			Deformation Requirements (mm)****		
		Diameter (mm)	Cross-Sectional Area (mm²)	Perimeter (mm)	Maximum Average Spacing	Minimum Average Height	Maximum Gap*****
10 [3]	0.560	9.5	71	29.9	6.7	0.39	3.6
13 [4]	0.994	12.7	129	39.9	8.9	0.51	4.9
16 [5]	1.552	15.9	199	49.9	11.1	0.71	6.1
19 [6]	2.235	19.1	284	59.8	13.3	0.97	7.3
22 [7]	3.042	22.2	387	69.8	15.5	1.12	8.5
25 [8]	3.973	25.4	510	79.8	17.8	1.27	9.7
29 [9]	5.059	28.7	645	90.0	20.1	1.42	10.9
32 [10]	6.404	32.3	819	101.3	22.6	1.63	12.4
36 [11]	7.907	35.8	1006	112.5	25.1	1.80	13.7
43 [14]	11.38	43.0	1452	135.1	30.1	2.16	16.5
57 [18]	20.24	57.3	2581	180.1	40.1	2.59	21.9

* Copyright ASTM, Printed with Permission.
** Bar numbers approximate the number of millimeters of the nominal diameter of the bars. [Bar numbers are based on the number of eighths of an inch of the nominal diameter of the bars.]
*** The nominal dimensions of a deformed bar are equivalent to those of a plain round bar having the same weight per meter as the deformed bar.
**** Requirements for protrusions on the surface of the bar.
***** Chord 12.5% of Nominal Perimeter.

Table 2.2 Types and Properties of Reinforcing Bars According to ASTM (Somayaji, 2001) (Reprinted by permission of Pearson Education, Inc.)

ASTM Designation	Type	Grade	Tensile strength min., MPa (ksi)	Yield strength* Min., MPa (ksi)	Size availability (No.)
A615	Billet steel bars (plain and deformed)	40 60 75	483 (70) 620 (90) 689 (100)	276 (40) 414 (60) 517 (75)	3~6 3~18 11~18
A616	Rail steel (plain and deformed)	50 60	552 (80) 620 (90)	345 (50) 474 (60)	3~11 3~11
A617	Axle steel (plain and deformed)	40 60	483 (70) 620 (90)	276 (40) 414 (60)	3~11 3~11
A706	Low-alloy steel deformed bars	60	552 (80)	414~538 (60~78)	3~18

* When the steel does not have a well-defined yield point, yield strength is the stress corresponding to a strain of 0.005 m/m (0.5% extension) for grades 40, 50, and 60, and a strain of 0.0035 m/m (0.35% extension) for grade 75 of A615, A616, and A617 steels. For A706 steel, grade point is determined at a strain of 0.0035 m/m.

Table 2.3 Required Properties for Seven-Wire Strand

Property	Stress-relieved		Low-relaxation (max. Percent)	
	Grade 250	Grade 270	Grade 250	Grade 270
Breaking strength,* MPa (ksi)	1725 (250)	1860 (270)	1725 (250)	1860 (270)
Yield strength (1% extension)	85% of breaking strength		90% of breaking strength	
Elongation (min. percent)	3.5		3.5	
Relaxation** (max. percent)				
Load=70% min. breaking strength	—		2.5	
Load=80% min. breaking strength	—		3.5	

* Breaking strength is the maximum load required to break one or more wires.

** Relaxation is the reduction in stress that occurs when a constant strain is applied over an extended time period. The specification is for a load duration of 1000 hours at a test temperature of 20 ± 2 °C (68 ± 3 °F).

The test specimen can be either full sized or machined into a shape, as prescribed in the product specifications for the material being tested. It is desirable to use a small cross-sectional area at the center portion of the specimen to ensure fracture within the gauge length. Several cross-sectional shapes are permitted, such as round and rectangular, as shown in Figure 2.2. Plate, sheet, round rod, wire, and tube specimens may be used. A 12.5 (1/2 in.) diameter round specimen is used in many cases. The gauge length over which the elongation is measured typically is four times the diameter for most round-rod specimens.

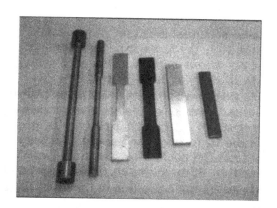

Figure 2.2 Tension test specimens with round and rectangular cross-sections.

Figure 2.3 Tension test on a round steel specimen showing grips and an extensometer with an LVDT.

Various types of gripping devices may be used to hold the specimen, depending on its shape. In all cases, the axis of the test specimen should be placed at the center of the testing machine head to ensure axial tensile stresses within the gauge length without bending. An extensometer with a dial gauge (Figure 1.24) or an LVDT (Figure 1.28) is used to measure the deformation of the entire gauge length. The test is performed by applying an axial load to the specimen at a specified rate. Figure 2.3 shows a tensile test being performed on a round steel specimen using an LVDT extensometer to measure the deformation.

As discussed in Chapter 1, mild steel has a unique stress-strain relation (Figure 2.4). Here, a linear elastic response is displayed up to the proportion limit. As the stress is increased beyond the proportion limit, the steel will yield, at which time the strain will increase without an increase in stress (actually the stress will slightly decrease). As tension increases past the yield point, strain increases following a nonlinear relation up to the point of failure.

Note that the decrease in stress after the peak does not mean a decrease in strength. In

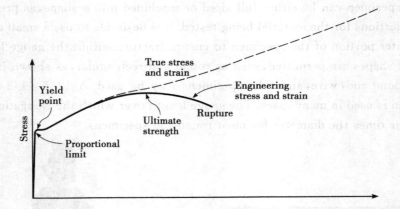

Figure 2.4 Typical stress-strain behavior of mild steel.

fact, the actual stress continues to increase until failure. The reason for the apparent decrease is that a neck is formed in the steel specimen, causing an appreciable decrease in the cross-sectional area. The traditional, or engineering, way of calculating the stress and strain uses the originalcross-sectional area and gauge length. If the stress and stains are calculated based on the instantaneous cross-sectional area and gauge length, a *true stress-strain curve* is obtained, which is different than the *engineering stress-strain curve* (Figure 2.4).

As shown in Figure 2.4, the true stress is larger than the engineering stress, because of the reduced cross-sectional area at the neck. Also, the true strain is larger than the engineering strain, since the increase in length at the vicinity of the neck is much larger than the increase in length outside of the neck. The specimen experiences the largest deformation (contraction of the cross-sectional area and increase in length) at the regions closest to the neck, due to the nonuniform distribution of the deformation. The large increase in length at the neck increases the true strain to a large extent because the definition of true strain utilizes a ratio of the change in length in an infinitesimal gauge length. By decreasing the gauge length toward an infinitesimal size and increasing the length due to localization in the neck, the numerator of an expression is increased while the denominator stays small, resulting in a significant increase in the ratio of the two numbers. Note that when calculating the true strain, a small gauge length should be used at the neck, since the properties of the material (such as the cross section) at the neck represent the true material properties. For various practical applications, however, the engineering stresses and strains are used, rather than the true stresses and strains.

Different carbon-content steels have different stress-strain relations. Increasing the carbon content in the steel increases the yield stress and reduces the ductility. Figure 2.5 shows the tension stress-strain diagram for hot-rolled steel bars containing carbons from 0.19% to 0.90%. Increasing the carbon content from 0.19% to 0.90% increases the yield stress from

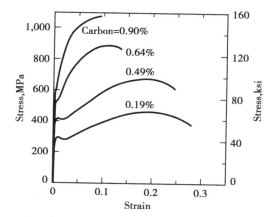

Figure 2.5 Tensile stress-strain diagrams of hot-rolled steel bars with different carbon contents.

280 MPa to 620 MPa (40 ksi to 90 ksi). Also, this increase in carbon content decreases the fracture strain from about 0.27 m/m to 0.09 m/m. Note that the increase in carbon content does not change the modulus of elasticity.

Sample Problem 2.1

A steel alloy bar 100 mm long with a rectangular cross section of 10 mm×40 mm is subjected to tension with a load of 89 kN and experiences an increase in length of 0.1 mm. If the increase in length is entirely elastic, calculate the modulus of elasticity of the steel alloy.

Solution

$$\sigma = \frac{89\,000}{0.01 \times 0.04} = 0.225 \times 10^9 \text{ Pa} = 0.2225 \text{ GPa}$$

$$\varepsilon = \frac{0.1}{100} = 0.001 \text{ mm/mm}$$

$$E = \frac{\sigma}{\varepsilon} = \frac{0.2225}{0.001} = 222.5 \text{ GPa}$$

Sample Problem 2.2

A steel specimen is tested in tension. The specimen is 1 in. wide by 0.5 in. thick in the test region. By monitoring the load dial of the testing machine, it was found that the specimen yielded at a load of 36 kips and fractured at 48 kips.
  a. Determine the tensile stresses at yield and at fracture.
  b. If the original gauge length was 4 in., estimate the gauge length when the specimen is stressed to 1/2 the yield stress.

Solution

a. Yield stress $(\sigma_y) = 36/(1 \times 0.5) = 72$ ksi
   Fracture stress $(\sigma_f) = 48/(1 \times 0.5) = 96$ ksi
b. Assume $E = 30 \times 10^6$ psi
   $\varepsilon = (1/2)\sigma_y/E = (1/2) \times 72 \times 10^3/(30 \times 10^6) = 0.0012$ in./in.
   $\Delta L = L\varepsilon = 4 \times 0.0012 = 0.0048$ in.
   Final gauge length $= 4 + 0.0048 = 4.0048$ in.

Steel is generally assumed to be a homogenous and isotropic material. However, in the production of structural members, the final shape may be obtained by cold rolling. This essentially causes the steel to undergo plastic deformations, with the degree of deformation varying throughout the member. As discussed in Chapter 1, plastic deformation causes an increase in yield strength and a reduction in ductility. Figure 2.6 demonstrates that the measured properties vary, depending on the orientation of the sample relative to the axis of rolling (Hassett, 2003). Thus, it is necessary to specify how the sample is collected when evaluating the mechanical properties of steel.

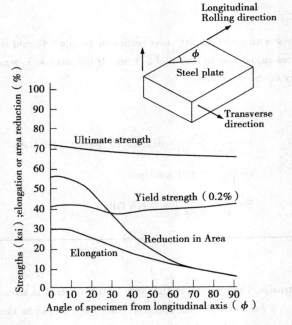

Figure 2.6 Example of effect of specimen orientation on measured tensile properties of steel.

2.2.2 Torsion Test

The torsion test (ASTM E143) is used to determine the shear modulus of structural materials. The shear modulus is used in the design of members subjected to torsion, such as rotating shafts and helical compression springs. In this test a cylindrical, or tubular, specimen is loaded either incrementally or continually by applying an external torque to cause a uniform twist within the gauge length (Figure 2.7). The amount of applied torque and the corresponding angle of twist are measured throughout the test. Figure 2.8 shows the shear stress-strain curve. The shear modulus is the ratio of maximum shear stress to the corresponding shear strain below the proportional limit of the material, which is the slope of the straight line between R (a pretorque stress) and P (the proportional limit). For a circular cross section, the maximum shear stress (τ_{max}), shear strain (γ), and the shear modulus (G) are determined by the equations

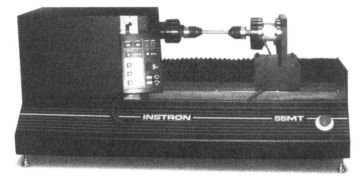

Figure 2.7 Torsion test apparatus.

$$\tau_{max} = \frac{Tr}{J} \quad (2.1)$$

$$\gamma = \frac{\theta r}{L} \quad (2.2)$$

$$G = \frac{\tau_{max}}{\gamma} = \frac{TL}{J\theta} \quad (2.3)$$

where

T = torque

r = radius

J = polar moment of inertia of the specimen about its center, $\pi r^4/2$ for a solid circular cross section.

θ = angle of twist in radians

L = gauge length

The test method is limited to materials and stresses at which creep is negligible compared with the strain produced immediately upon loading. The test specimen should be sound, without imperfections near the surface. Also, the specimen should be straight and of uniform diameter for a length equal to the gauge length plus two to four diameters. The gauge length should be at least four diameters. During the test, torque is read from a dial gauge or a readout device attached to the testing machine, while the angle of twist may be measured using a torsiometer fastened to the specimen at the two ends of the gauge length. A curve-fitting procedure can be used to estimate the straight-line portion of the shear stress-strain relation of Figure 2.8 (ASTM E143).

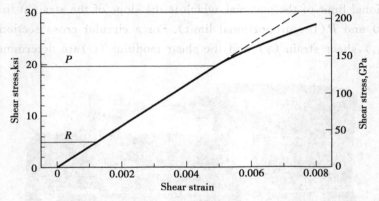

Figure 2.8 Typical shear stress-strain diagram of steel (ASTME143). Copyright ASTM. Reprinted with permission.

Sample Problem 2.3

A rod with a length of 1 m and a radius of 20 mm is made of high-strength steel. The rod is subjected to a torque T, which produces a shear stress below the proportional limit. If the cross section at one end is rotated 45 degrees in relation to the other end, and the shear modulus G of the material is 90 GPa, what is the amount of applied torque?

Solution

$$J = \pi r^4/2 = \pi (0.02)^4/2 = 0.2513 \times 10^{-6} \text{ m}^4$$
$$\theta = 45(\pi/180) = \pi/4$$
$$T = GJ\theta/L = (90 \times 10^9) \times (0.2513 \times 10^{-6}) \times (\pi/4)/1$$
$$= 17.8 \times 10^3 \text{ N} \cdot \text{m} = 17.8 \text{ kN} \cdot \text{m}$$

2.2.3 Charpy V Notch Impact Test

The Charpy V Notch impact test (ASTM E23) is used to measure the toughness of the material or the energy required to fracture a V-notched simply supported specimen. The test is used for structural steels in tension members.

Figure 2.9 Charpy V notch specimens.

Figure 2.10 Charpy V notch impact testing machine.

The standard specimen is 55 mm×10 mm×10 mm with a V notch at the center, as shown in Figure 2.9. Before testing, the specimen is brought to the specified temperature for a minimum of 5 min in a liquid bath or 30 min in a gas medium. The specimen is inserted into the Charpy V notch impact-testing machine (Figure 2.10) using centering tongs. The swinging arm of the machine has a striking tip that impacts the specimen on the side opposite the V notch. The striking head is released from the pretest position, striking and fracturing the specimen. By fracturing the test specimen, some of the kinetic energy of the striking head is absorbed, thereby reducing the ultimate height the strike head attains. By measuring the height the strike head attains after striking the specimen, the energy required to fracture the specimen is computed. This energy is measured in m · N as indicated on a gauge attached to the machine.

The lateral expansion of the specimen is typically measured after the test using a dial gauge device. The lateral expansion is a measure of the plastic deformation during the test. The higher the toughness of the steel, the larger the lateral expansion.

Figure 1.13, in Chapter 1, shows the typical energy (toughness) required to fracture structural steel specimens at different temperatures. The figure shows that the required energy is high at high temperatures and low at low temperatures. This indicates that the material changes from ductile to brittle as the temperature decreases.

The fracture surface typically consists of a dull shear area (ductile) at the edges and a shiny cleavage area (brittle) at the center, as depicted in Figure 2.11. As the toughness of the steel decreases, due to lowering the temperature, for example, the shear area decreases while the cleavage area increases.

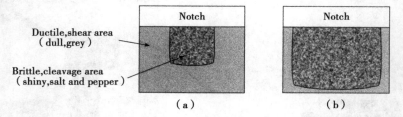

Figure 2.11 Fracture surface of Charpy V notch specimen: (a) at high temperature and (b) at low temperature.

Sample Problem 2.4

A Charpy V Notch (CVN) test was performed on a steel specimen and produced the following readings:

Temperature (°F)	Toughness (ft. lb)
−40	5
30	7
100	28
170	66
240	79
310	80

Plot the toughness-versus-temperature relation, and determine the temperature transition zone between ductile and brittle behavior.

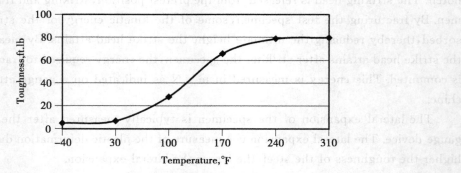

Figure SP2.4

Solution

The toughness-versus-temperature relation is as shown in Figure SP2.4. From the figure, the temperature transition zone between ductile and brittle behavior can be seen to be 30 to 240 °F.

2.2.4 Bend Test

In many engineering applications, steel is bent to a desired shape, especially in the case of reinforcing steel. The ductility to accommodate bending is checked by performing the semiguided bend test (ASTM E290). The test evaluates the ability of steel, or a weld, to resist cracking during bending. The test is conducted by bending the specimen through a specified angle and to a specified inside radius of curvature. When complete fracture does not occur, the criterion for failure is the number and size of cracks found on the tension surface of the specimen after bending.

The bend test is made by applying a transverse force to the specimen in the portion that is being bent, usually at midlength. Three arrangements can be used, as illustrated in Figure 2.12. In the first arrangement, the specimen is fixed at one end and bent around a reaction pin or mandrel by applying a force near the free end, as shown in Figure 2.12(a). In the second arrangement, the specimen is held at one end and a rotating device is used to bend the specimen around the pin or mandrel, as shown in Figure 2.12(b). In the third arrangement, a force is applied in the middle of a specimen simply supported at both ends, Figure 2.12(c).

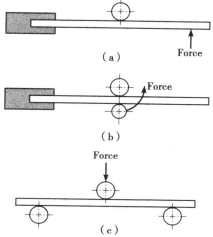

Figure 2.12 Schematic fixtures for semi-guided bend test (ASTM E290). Copyright ASTM. Reprinted with permission.

2.2.5 Hardness Test

Hardness is a measure of a material's resistance to localized plastic deformation, such as a small dent or scratch on the surface of the material. A certain hardness is required for many machine parts and tools. Several tests are available to evaluate the hardness of materials. In these tests an indenter (penetrator) is forced into the surface of the material with a specified load magnitude and rate of application. The depth, or the size, of the indentation is measured and related to a hardness index number. Hard materials result in small impressions, corresponding to high hardness numbers. Hardness measurements depend on test conditions and are, therefore, relative. Correlations and tables are available to convert the hardness measurements from one test to another and to approximate the tensile strength of the material (ASTM A370).

One of the methods commonly used to measure hardness of steel and other metals is the Rockwell hardness test (ASTM E18). In this test the depth of penetration of a diamond cone, or a steel ball, into the specimen is determined under fixed conditions (Figure 2.13). A preliminary load of 10 kg is applied first, followed by an additional load. The Rockwell number, which is proportional to the difference in penetration between the preliminary and total loads, is read from the machine by means of a dial, digital display, pointer, or other device. Two scales are frequently used, namely, B and C. Scale B uses a 1.588 mm steel ball indenter and a total load of 100 kg, while scale C uses a diamond spheroconical indenter with a 120° angle and a total load of 150 kg.

Figure 2.13 Rockwell hardness test machine.

To test very thin steel or thin surface layers, the Rockwell superficial hardness test is used. The procedure is the same as the Rockwell hardness test except that smaller preliminary and total loads are used. The Rockwell hardness number is reported as a number, followed by the symbol HR, and another symbol representing the indenter and forces used. For example, 68 HRC indicates a Rockwell hardness number of 68 on Rockwell C scale.

Hardness tests are simple, inexpensive, nondestructive, and do not require special specimens. In addition, other mechanical properties, such as the tensile strength, can be estimated from the hardness numbers. Therefore, hardness tests are very common and are typically performed more frequently than other mechanical tests.

2.2.6 Ultrasonic Testing

Ultrasonic testing is a nondestructive method for detecting flaws in materials. It is particularly useful for the evaluation of welds. During the test, a sound wave is directed toward the weld joint and reflected back from a discontinuity, as shown on Figure 2.14. A sensor captures the energy of the reflected wave and the results are displayed on an oscilloscope. This method is highly sensitive in detecting planar defects, such as incomplete weld fusion, delamination, or cracks (Hassett, 2003).

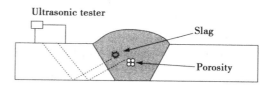

Figure 2.14 Ultrasonic test of welds.

2.3 Steel Corrosion

Corrosion is defined as the destruction of a material by electrochemical reaction to the environment. For simplicity, corrosion of steel can be defined as the destruction that can be detected by rust formation. Corrosion of steel structures can cause serious problems and embarrassing and/or dangerous failures. For example, corrosion of steel bridges, if left unchecked, may result in lowering weight limits, costly steel replacement, or collapse of the structure. Other examples include corrosion of steel pipes, trusses, frames, and other structures. It is estimated that the cost of corrosion in the United States alone is $ 8 billion each year (Frank and Smith 1990).

Corrosion is an electrochemical process; that is, it is a chemical reaction in which there is transfer of electrons from one chemical species to another. In the case of steel, the transfer is between iron and oxygen, a process called oxidation reduction. Corrosion requires the following four elements (without any of them corrosion will not occur):

1. an anode—the electrode where corrosion occurs
2. a cathode—the other electrode needed to form a corrosion cell
3. a conductor—a metallic pathway for electrons to flow
4. an electrolyte—a liquid that can support the flow of electrons

Steel, being a heterogeneous material, contains anodes and cathodes. Steel is also an electrical conductor. Therefore, steel contains three of the four elements needed for corrosion, while moisture is usually the fourth element (electrolyte).

The actual electrochemical reactions that occur when steel corrodes are very complex.

However, the basic reactions for atmospherically exposed steel in a chemically neutral environment are dissolution of the metal at the anode and reduction of oxygen at the cathode.

Contaminants deposited on the steel surface affect the corrosion reactions and the rate of corrosion. Salt, from deicing or a marine environment, is a common contaminant that accelerates corrosion of steel bridges and reinforcing steel in concrete.

The environment plays an important role in determining corrosion rates. Since an electrolyte is needed in the corrosion reaction, the amount of time the steel stays wet will affect the rate of corrosion. Also, contaminants in the air, such as oxides or sulfur, accelerate corrosion. Thus, areas with acid rain, coal-burning power plants, and other chemical plants may accelerate corrosion.

Since steel contains three of the four elements needed for corrosion, protective coatings can be used to isolate the steel from moisture, the fourth element. There are three mechanisms by which coatings provide corrosion protection (Hare 1987):

1. Barrier coatings work solely by isolating the steel from the moisture. These coatings have low water and oxygen permeability.
2. *Inhibitive primer coatings* contain passivating pigments. They are lowsolubility pigments that migrate to the steel surface when moisture passes through the film to passivate the steel surface.
3. Sacrificial primers (cathodic *protection*) contain pigments such as elemental zinc. Since zinc is higher than iron in the galvanic series, when corrosion conditions exist the zinc gives up electrons to the steel, becomes the anode, and corrodes to protect the steel. There should be close contact between the steel and the sacrificial primer in order to have an effective corrosion protection.

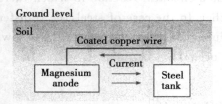

Figure 2.15 Cathodic protection of an underground pipeline using a magnesium sacrificial anode.

Cathodic protection can take forms other than coating. For example, steel structures such as water heaters, underground tanks and pipes, and marine equipment, can be electrically connected to another metal that is more reactive in the particular environment, such as magnesium or zinc. Such reactive metal (sacrificial anode) experiences oxidation and gives up electrons to the steel, protecting the steel from corrosion. Figure 2.15 illustrates an underground steel tank that is electrically connected to a magnesium sacrificial anode (Fontana and Green 1978).

SUMMARY

The history of civil engineering is closely tied to that of steel, and this will continue into the foreseeable future. With the development of modern production facilities, the availability of a wide variety of economical steel products is virtually assured. High strength, ductility, the ability to carry tensile as well as compressive loads, and the ability to join members either with welding or mechanical fastening are the primary positive attributes of steel as a structural material. The properties of steel can be tailored to meet the needs of specific applications through alloying and heat treatments. The primary shortcoming of steel is its tendency to corrode. When using steel in structures, the engineer should consider the means for protecting the steel from corrosion over the life of the structure.

QUESTIONS AND PROBLEMS

2.1 What is the chemical composition of steel? What is the effect of carbon on the mechanical properties of steel?

2.2 Why does the iron-carbon phase diagram go only to 6.7% carbon?

2.3 Draw a simple iron-carbon phase diagram showing the liquid, liquid-solid, and solid phases.

2.4 What is the typical maximum percent of carbon in steel used for structures?

2.5 What are the typical uses of structural steel?

2.6 Why is reinforcing steel used in concrete? Discuss the typical properties of reinforcing steel.

2.7 Name three mechanical tests used to measure properties of steel.

2.8 The following laboratory tests are performed on steel specimens:
   a. Tension test
   b. Charpy V notch test
   c. Bend test
   What are the significance and use of these tests?

References

Bouchard, S., and G. Axmann. ASTM A913 Grades 50 and 65: Steels for Seismic Applications. *STESSA Conference Proceeding*, 2000.

Budinski K. G. *Engineering Materials, Properties and Selection*. 5th ed. Upper Saddle River, NJ: Prentice Hall,

1996.

Callister, W. D., Jr. *Materials Science and Engineering—an Introduction*. 6th ed. New York: John Wiley and Sons, 2003.

Carter, C. J., Are Yor Properly Specifying Materials? *Modern Steel Construction*, January edition, 2004.

Cordon, W. A. *Properties, Evaluation, and Control of Engineering Materials*. New York: McGraw-Hill, 1979.

Federal Highway Administration. A High Performance Steel Scorecard, Focus, Http://www.tfhrc.gov/focus/apr02/steel.htm, 2002.

Fontana, M. G. and N. D. Greene. *Corrosion Engineering*. 2d ed. New York: McGraw-Hill, 1978.

Frank, K. H. and L. M. Smith. *Highway Materials Engineering: Steel, Welding and Coatings*. Publication No. FHWA-HI-90-006. Washington, DC: Federal Highway Administration, 1990.

Hare, C. H. *Protective Coatings for Bridge Steel*. National Cooperative Highway Research Program, Synthesis of Highway Practice No. 136. Washington, DC: Transportation Research Board, 1987.

Hassett, P. M. *Steel Construction in the New Millennium*. Hassett Engineering, Inc., Castro Valley, CA, 2003.

CHAPTER 3
ALUMINUM

Aluminum is the most plentiful metal on Earth, representing 8% of its crust. Although plentiful, aluminum exists primarily as oxides. The process of extracting aluminum from oxide is very energy intensive. In fact, approximately 2% to 3% of the electricity used in the United States is consumed in aluminum production. This high energy requirement makes recycling of aluminum products economical. Of the 24 million tons of aluminum produced annually, approximately 75% is from ore reduction and 25% is from recycled materials.

The properties of pure aluminum are not suitable for structural applications. Some industrial applications require pure aluminum, but otherwise, alloying elements are almost always added. These alloying elements, along with cold working and heat treatments, impart characteristics to the aluminum that make this product suitable for a wide range of applications. Here, the term aluminum is used to refer to both the pure element and to alloys.

In terms of the amount of metal produced, aluminum is second only to steel. About 25% of aluminum produced is used for containers and packaging, 20% for architectural applications, such as doors, windows, and siding, and 10% for electrical conductors. The balance is used for industrial goods, consumer products, aircraft, and highway vehicles.

Aluminum accounts for 80% of the structural weight of aircraft, and its use in the automobile and light truck industry has increased 300% since 1971 (Reynolds Metals Company 1996). However, use of aluminum for infrastructure applications has been limited. Of the approximately 600,000 bridges in the United States, only nine have primary structural members made of aluminum. Two reasons for the limited use of aluminum are the relatively high initial cost when compared with steel and the lack of performance information on aluminum structures.

Aluminum has many favorable characteristics and a wide variety of applications. The advantages of aluminum are that it (Budinski 1996)

- has one-third the density of steel
- has good thermal and electrical conductivity
- has high strength-to-weight ratio
- can be given a hard surface by anodizing and hard coating
- has alloys that are weldable
- will not rust
- has high reflectivity
- can be die cast
- is easily machined
- has good formability
- is nonmagnetic
- is nontoxic

Aluminum's high strength-to-weight ratio and its ability to resist corrosion are the primary factors that make aluminum an attractive structural engineering material. Although aluminum alloys can be formulated with strengths similar to steel products, the modulus of elasticity of aluminum is only about one-third that of steel. Thus, the dimensions of structural elements must be increased to compensate for the lower modulus of elasticity of aluminum.

3.1 Aluminum Production

Aluminum production uses processes that were developed in the 1880s. Bayer developed the sodium aluminate leaching process to produce pure alumina (Al_2O_3). Hall and Héroult, working independently, developed an electrolytic process for reducing the alumina to pure aluminum. The essence of the aluminum production process is shown in Figure 3.1.

The production of aluminum starts with the mining of the aluminum ore, bauxite. Commercial grade bauxite contains between 45% and 60% alumina. The bauxite is crushed, washed to remove clay and silica materials, and is kiln dried to remove most of the water. The crushed bauxite is mixed with soda ash and lime and passed through a digester, pressure reducer, and settling tank to produce a concentrated solution of sodium aluminate. This step removes silica, iron oxide, and other impurities from the sodium aluminate solution. The solution is seeded with hydrated alumina crystals in precipitator towers. The seeds attract other alumina crystals and form groups that are heavy enough to settle out of solution. The alumina hydrate crystals

Chapter 3 Aluminum   65

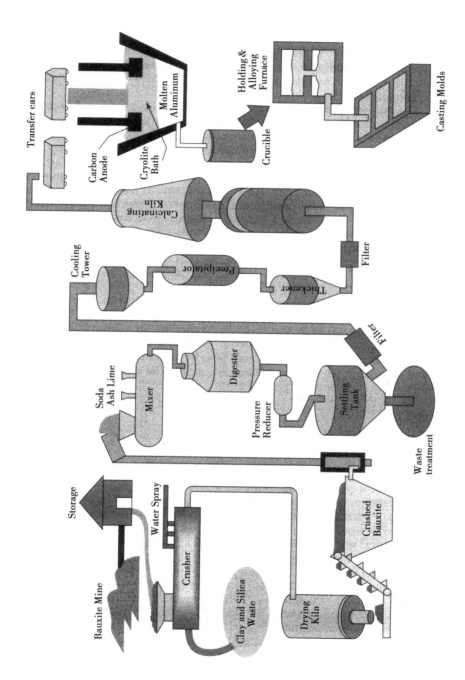

Figure 3.1 Aluminum production process.

are washed to remove remaining traces of impurities and are calcined in kilns to remove all water. The resulting alumina is ready to be reduced with the Hall-Héroult process. The alumina is melted in a cryolite bath (a molten salt of sodium-aluminum-fluoride). An electric current is passed between anodes and cathodes of carbon to separate the aluminum and oxygen molecules. The molten aluminum is collected at the cathode at the bottom of the bath. The molten aluminum, with better than 99% purity, is siphoned off to a crucible. It is then processed in a holding furnace. Hot gases are passed through the molten material to further remove any remaining impurities. Alloying elements are then added.

The molten aluminum is either shipped to a foundry for casting into finished products or is cast into ingots. The ingots are formed by a directchill process that produces huge sheets for rolling mills, round loglike billets for extrusion presses, or square billets for production of wire, rod, and bar stock.

Final products are made by either casting, which is the oldest process, or deforming solid aluminum stock. Three forms of casting are used: die casting, permanent mold casting, and sand casting. The basic deformation processes are forging, impact extrusion, stamping, drawing, and drawing plus ironing. Many structural shapes are made with the extrusion process. Either cast or deformed products can be machined to produce the final shape and surface texture, and they can be heat treated to alter the mechanical behavior of the aluminum. Casting and forming methods are summarized in Table 3.1.

Table 3.1 Casting and Forming Methods for Aluminum Products
(Extracted from Reynolds Infrastructure, 1996)

Casting Methods	
Sand Casting	Sand with a binder is packed around a pattern. The pattern is removed and molten aluminum is poured in, reproducing the shape. Produces a rough texture which can be machined or otherwise surfaced if desired. Economical for low volume production and for making very large parts. Also applicable when an internal void must be formed in the product.
Permanent Mold Casting	Molten aluminum is poured into a reusable metal mold. Economical for large volume production.
Die Casting	Molten aluminum is forced into a permanent mold under high pressure. Suitable for mass production of precisely formed castings.
Forming Methods	
Extrusion	Aluminum heated to 425 to 540 °C (800 to 1000 °F) is forced through a die. Complex cross sections are possible, including incompletely or completely enclosed voids. A variety of architectural and structural members are formed by extrusion, including tubes, pipes, I-beams, and decorative components, such as window and door frames.
Rolling	Rollers compress and elongate heated aluminum ingots, producing plates (more than 6 mm (0.25 in) thick), sheets (0.15 to 6 mm) (0.006 to 0.25 in.) thick, and foil (less than 0.15 mm (0.006 in.)).

Continued

Roll Forming	Shaping of sheet aluminum by passing stock between a series of special rollers, usually in stages. Used for mass production of architectural products, such as moldings, gutters, downspouts, roofing, siding and frames for windows and screens.
Brake Forming	Forming of sheet products with a brake press. Uses simpler tooling than roll forming but production rates are lower and the size of the product is limited.
Cutting Operations	Production of outline shapes by blanking and cutting. In blanking, a punch with the desired shape is pressed through a matching die. Used for mass production of flat shapes. Holes through a sheet are produced by piercing and perforating. Stacks of sheets can be trimmed or cut to an outline shape by a router or sheared in a guillotine-action shear.
Embossing	Shaping an aluminum sheet by pressing between mated rollers or dies, producing a raised pattern on one side and its negative indent on the other side.
Drawing	Shaping an aluminum sheet by drawing it through the gap between two mated dies in a press.
Superplastic Forming	An aluminum sheet is heated and forced over or into a mold by air pressure. Complex and deep contour shapes can be produced but the process is slow.

When recycling aluminum, the scrap stock is melted in a furnace. The molten aluminum is purified and alloys are added. This process takes only about 5% of the electricity that is needed to produce aluminum from bauxite.

In addition to these conventional processes, very high strength aluminum parts can be produced using powder metallurgy methods. A powdered aluminum alloy is compacted in a mold. The material is heated to a temperature that fuses the particles into a unified solid.

3.2 Aluminum Testing and Properties

Typical properties are provided in Tables 3.2 and 3.3 for non-heat-treatable and heat-treatable wrought aluminum alloys, respectively. Typical properties for cast aluminum alloys that may be used for structural applications are given in Table 3.4. These values are only an indication of the properties of cast aluminum alloys. Material properties of cast members can vary throughout the body of the casting due to differential cooling rates.

Table 3.2 Properties of Select Non-Heat-Treatable Wrought Aluminum Alloys

Alloy		Tension				Elongation[1] (thickness)		Hardness[2]	Shear Ultimate		Fatigue[3] Endurance Limit		Nominal Chemical Composition
		Ultimate		Yield		1/16″	1/2″		ksi	MPa	ksi	MPa	
		ksi	MPa	ksi	MPa								
1060	O	10	69	4	28	43		19	7	48	3	21	99.6 Al
	H-12	12	83	11	76	16		23	8	55	4	28	
	H-14	14	97	13	90	12		26	9	62	5	34	
	H-16	16	110	15	103	5		30	10	69	6.5	45	
	H-18	19	131	18	124	6		35	11	76	6.5	45	
1100	O	13	90	5	34	35	45	23	9	62	5	34	99. Al
	H-12	16	110	15	103	12	25	28	10	69	6	41	
	H-14	18	124	17	117	9	50	32	11	76	7	48	
	H-16	21	145	20	138	6	17	38	12	83	9	62	
	H-18	24	165	22	152	5	15	44	13	90	9	62	
3003	O	16	110	6	41	30	40	28	11	76	7	48	1.2 Mn
	H-12	19	131	18	124	10	20	35	12	83	8	55	
	H-14	22	152	21	145	8	16	40	14	97	9	62	
	H-16	26	179	25	172	5	14	47	15	103	10	69	
	H-18	29	200	27	186	4	10	55	16	110	10	69	

Chapter 3 Aluminum 69

Alloy	Temper											Composition
5005	O	18	124	6	41	25		28	11	76			0.8 Mg
	H-12	20	138	19	131	10			14	97			
	H-14	23	159	22	152	6			14	97			
	H-16	26	179	25	172	5			15	103			
	H-18	29	200	28	193	4			16	110			
	H-32	20	138	17	117	11		36	14	97			
	H-34	23	159	20	138	8		41	14	97			
	H-36	26	179	24	165	6		46	15	103			
	H-38	29	200	27	186	5		51	16	110			
5086	O	38	262	17	117	22	30	60	23	159	21	145	Mg
	H-32	42	290	30	207	12	16	72	25	172	22	152	
	H-34	47	324	37	255	10	14	82	27	186	23	159	0.45 Mn
	H-111| 40	276	27	186		17	65	23	159	21	145	
	H-112| 39	269	19	131	14		64	23	159	21	145	
	H-116| 42	290	30	207		16	72	25	173	22	152	
5456	O	45	310	23	159	20	24	70	27	186	22	152	5.1 Mg
	H-111| 47	324	33	228		18	75	27	186	24	165	0.7 Mn
	H-112| 45	310	24	156		22	70	27	186			0.12 Cr
	H-116| 51	352	37	255		16	90	30	207	23	159	

[1] percent elongation over 2 in.
[2] Brinell number, 500-kg load
[3] 500,000,000 cycles of complete stress reversal using R. R. Moore type of machine and specimen.

Table 3.3 Properties of select Heat-Treatable Wrought Aluminum Alloys

Alloy		Tension				Elongation[1] (thickness)		Hardness[2]	Shear Ultimate		Fatigue[3] Endurance Limit		Nominal Chemical Composition
		Ultimate		Yield									
		ksi	MPa	ksi	MPa	1/16"	1/2"		ksi	MPa	ksi	MPa	
2014	O	27	186	14	97		18	45	18	124	13	0	4.5 Cu, 0.8 Mn
	T4/T451	62	427	42	290		50	105	38	262	20	138	0.8 Si, 0.4 M
	T6/T651	70	483	60	414		13	135	42	290	18	124	
6053	O	16	110	8	55		35	26	11	76	8	55	1.2 Mg
	T6	37	255	32	221		13	80	23	159	12	90	0.25 CR
6061	O	18	124	8	55	25	30	30	12	83	9	62	1.0 Mg, 0.6 SI
	T4/T451	35	241	21	145	22	25	65	24	165	14	97	0.25 Cu,
	T6/T651	45	310	40	276	12	17	95	30	207	14	97	0.25 Cr
	O	13	90	7	48			25	10	69	8	55	0.7 Mg
	T1	22	152	13	90	20	33	42	14	97	9	62	0.4 Si
	T4	25	172	13	90	22			16	110			
6063	T5	27	186	21	145	12	22	60	17	117	10	69	
	T6	35	241	31	214	12	18	73	22	152	10	69	
	T83	37	255	35	241	9		82	22	152			
	T831	30	207	27	186	10		70	18	124			
	T832	42	290	39	269	12		95	27	186			
7178	O	33	228	15	103	15	16	60	22	152			6.8 Zn, 2.0 Cu
	T6/T651	88	607	78	538	10	11	160	52	359	22	152	2.7 Mg, 0.3 Mn
	T76/T765	83	572	73	503		11						

[1]Percent elongation over 2 in.
[2]Brinell number, 500-kg load
[3]500,000,000 cycles of complete stress reversal using R. R. Moore type of machine and specimen.

Tests performed on aluminum are similar to those described for steel. These typically include stress-strain tensile tests to determine elastic modulus, yield strength, ultimate strength, and percent elongation. In contrast to steel, aluminum alloys do not display an upper and lower yield point. Instead, the stress-strain curve is linear up to the proportional limit, and then is a smooth curve up to the ultimate strength. Yield strength is defined based on the 0.20% strain offset method, as shown in Figure 3.2. As indicated earlier, the modulus of elasticity of aluminum alloys is on the order of 69 GPa and is not very sensitive to types of alloys or temper treatments.

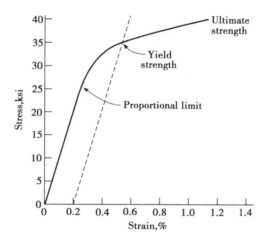

Figure 3.2 Aluminum stress-strain diagram

Sample Problem 3.1

An aluminum alloy rod with 10 mm diameter is subjected to a 5-kN tensile load. After the load was applied, the diameter was measured and found to be 9.997 mm. If the yield strength is 139 MPa, calculate the Poisson's ratio of the material.

Solution

$$\sigma = \frac{5000}{\pi d^2/4} = 63.7 \times 10^6 \text{ Pa} = 63.7 \text{ Mpa}$$

It is clear that the applied stress is well below the yield stress and, as a result, the deformation is elastic. Hence, assume that

$E = 69$ GPa

$$\varepsilon_{axial} = \frac{\sigma}{E} = \frac{63.7 \times 10^6}{69 \times 10^9} = 0.000923 \text{ m/m}$$

$$\Delta d = 9.997 - 10.000 = 0.003 \text{ m}$$

$$\varepsilon_{\text{lateral}} = \frac{-0.003}{10.000} = -0.0003 \text{ m/m}$$

$$v = \frac{-\varepsilon_{\text{lateral}}}{\varepsilon_{\text{axial}}} = \frac{0.0003}{0.000923} = 0.33$$

Table 3.4 Typical Properties of Select Cast Aluminum Alloys

Cast Alloy Designation	Tension					Hardness[2]	Shear Ultimate		Fatigure[3] Endurance Limit	
	Ultimate		Yield		Elongation[1]					
	ksi	MPa	ksi	MPa			ksi	MPa	ksi	MPa
356.0-T6[4]	40	276	27	186	5	90	32	221	13	90
356.0-T7[4]	33	228	24	165	5	70	25	172	11	76
A356.0-T61[4]	41	283	30	207	10	80				
A357.0-T6[4]	50	345	40	276	10	85	43	296	16	110
A444.0-T4[4]	23	159	10	69	21	45				
356.0-T6[5]	33	228	24	165	3.5	70	26	179	8.5	59
356.0-T7[5]	34	234	30	207	2.0	75	24	165	9.0	62
Almag 35 535.0[5]	40	276	21	145	13	70	28	193	10	69

[1] Percent elongation over 2 in.
[2] Brinell number, 500-kg load
[3] 500,000,000 cycles of complete stress reversal using R. R. Moore type of machine and specimen.
[4] Permanent mold
[5] Sand casting

Aluminum's coefficient of thermal expansion is 0.000023/°C, about twice as large as that of steel and concrete. Thus, joints between aluminum and steel or concrete must be designed to accommodate the differential movement.

Strengths of aluminum are considerably affected by temperature, as shown in Figure 3.3. At temperatures above 150 °C, tensile strengths are reduced considerably. The temperature at which the reduction begins and the extent of the reduction depends on the alloy. At temperatures below room temperature, aluminum becomes stronger and tougher as the temperature decreases.

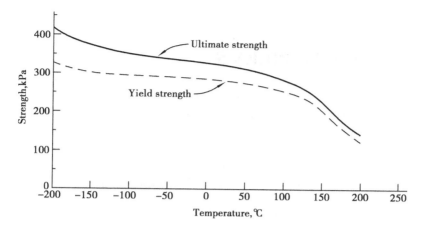

Figure 3.3 Tensile strength of aluminum at different temperatures. (Courtesy of the Aluminum Association, 1987.)

3.3 Corrosion

Aluminum develops a thin oxidation layer immediately upon exposure to the atmosphere. This tough oxide film protects the surface from further oxidation. The alloying elements alter the corrosion resistance of the aluminum. The alloys used for airplanes are usually given extra protection by painting or "cladding" with a thin coat of a corrosion-resistant alloy. Painting is generally not needed for medium-strength alloys used for structural applications.

Galvanic corrosion occurs when aluminum is in contact with any of several metals in the presence of an electrical conductor, such as water. The best protection for this problem is to break the path of the galvanic cell by painting, using an insulator, or keeping the dissimilar metals dry.

SUMMARY

Although aluminum has many desirable attributes, its use as a structural material in civil engineering has been limited, primarily by economic considerations and a lack of performance information. Aluminum alloys and heat treatments provide products with a wide range of characteristics. The advantages of aluminum relative to steel include lightweight, high strength-to-weight ratio, and corrosion resistance.

QUESTIONS AND PROBLEMS

3.1 Name the two primary factors that make aluminum an attractive structural engineering material.

3.2 Compare the strength and modulus of elasticity of aluminum alloys with those of steel.

3.3 An aluminum alloy specimen with a radius of 0.28 in. was subjected to tension until fracture and produced results shown in Table P3.3.

Table P3.3

Stress, ksi	Strain, 10^{-3} in./in.
8	0.6
17	1.5
27	2.4
35	3.2
43	4.0
50	4.6
58	5.2
62	5.8
64	6.2
65	6.5
67	7.3
68	8.1
70	9.7

a. Using a spreadsheet program, plot the stress-strain relationship.
b. Calculate the modulus of elasticity of the aluminum alloy.
c. Determine the proportional limit.
d. What is the maximum load if the stress in the bar is not to exceed the proportional limit?
e. Determine the 0.2% offset yield strength.
f. Determine the tensile strength.
g. Determine the percent of elongation at failure.

3.4 Discuss galvanic corrosion of aluminum. How can aluminum be protected from galvanic corrosion?

References

Aluminum Association. *Structural Design with Aluminum*. Washington, DC: The Aluminum Association, 1987.

Aluminum Association. *Aluminum Standards and Data*, 1993. Washington, DC: The Aluminum Association, 1993.

Budinski, K. G. *Engineering Materials, Properties and Selection*. 5th ed. Upper Saddle River, NJ: Prentice Hall, 1996.

Reynolds Metals Company. *Reynolds Infrastructure*. Richmond, VA: Reynolds Metals Company, 1996.

CHAPTER 4
AGGREGATES

There are two main uses of aggregates in civil engineering: as an underlying material for foundations and pavements and as ingredients in portland cement and asphalt concretes. By dictionary definition, aggregates are a combination of distinct parts gathered into a mass or a whole. Generally, in civil engineering the term aggregate means a mass of crushed stone, gravel, sand, etc., predominantly composed of individual particles, but in some cases including clays and silts. The largest particle size in aggregates may have a diameter as large as 150 mm and the smallest particle can be as fine as 5 to 10 microns. The balance of this chapter presents information about aggregates as used in construction. Information is not presented about the characteristics and properties of soils, as this is the purview of textbooks on geotechnical engineering.

4.1 Aggregate Sources

Natural sources for aggregates include gravel pits, river run deposits, and rock quarries. Generally, gravel comes from pits and river deposits, whereas crushed stones are the result of processing rocks from quarries. Usually, gravel deposits must also be crushed to obtain the needed size distribution, shape, and texture.

Manufactured aggregates can use slag waste from steel mills and expanded shale and clays to produce lightweight aggregates. Heavyweight concrete, used for radiation shields, can use steel slugs and bearings for the aggregate. Styrofoam beads can be used as an aggregate in lightweight concrete used for insulation.

4.2 Geological Classification

All natural aggregates result from the breakdown of large rock masses. Geologists classify rocks into three basic types: *igneous*, *sedimentary*, and *metamorphic*. Volcanic action produces igneous rocks by hardening or crystallizing molten material, magma. The magma cools either at the earth's surface, when it is exposed to air or water, or within the crust of the earth. Cooling at the surface produces extrusive igneous rocks, while cooling underground produces *intrusive* igneous rocks. In general, the extrusive rocks cool much more rapidly than the intrusive rocks. Therefore, we would expect extrusive igneous rocks to have a fine grain size and potentially to include air voids and other inclusions. Intrusive igneous rocks have larger grain sizes and fewer flaws. Igneous rocks are classified based on grain size and composition. Coarse grains are larger than 2 mm and fine grains are less than 0.2 mm. Classification based on composition is a function of the silica content, specific gravity, color, and the presence of free quartz.

Sedimentary rocks coalesce from deposits of disintegrated existing rocks or inorganic remains of marine animals. Wind, water, glaciers, or direct chemical precipitation transport and deposit layers of material that become sedimentary rocks, resulting in a stratified structure. Natural cementing binds the particles together. Classification is based on the predominant mineral present: calcareous (limestone, chalk, etc.), siliceous (chert, sandstone, etc.), and argillaceous (shale, etc.).

Metamorphic rocks form from igneous or sedimentary rocks that are drawn back into the earth's crust and exposed to heat and pressure, re-forming the grain structure. Metamorphic rocks generally have a crystalline structure, with grain sizes ranging from fine to coarse.

All three classes of rock are used successfully in civil engineering applications. The suitability of aggregates from a given source must be evaluated by a combination of tests to check physical, chemical, and mechanical properties, and must be supplemented by mineralogical examination. The best possible prediction of aggregate suitability for a given application is that based on historical performance in a similar design.

4.3 Evaluation of Aggregate Sources

Civil engineers select aggregates for their ability to meet specific project requirements, rather than their geologic history. The physical and chemical properties of the rocks determine the acceptability of an aggregate source for a construction project. These characteristics vary within a quarry or gravel pit, making it necessary to continually sample and test the materials as the aggregates are being produced.

Due to the quantity of aggregates required for a typical civil engineering application, the cost and availability of the aggregates are important when selecting an aggregate source. Frequently, one of the primary challenges facing the materials engineer on a project is how to use the locally available material in the most cost-effective manner.

Potential aggregate sources are usually evaluated for quality of the larger pieces, the nature and amount of fine material, and the gradation of the aggregate. The extent and quality of rock in the quarry is usually investigated by drilling cores and performing trial blasts (or shots) to evaluate how the rock breaks and by crushing some materials in the laboratory to evaluate grading, particle shape, soundness, durability, and amount of fine material. Cores are examined petrographically for general quality, suitability for various uses, and amount of deleterious materials. Potential sand and gravel pits are evaluated by collecting samples and performing sieve analysis tests. The amount of large gravel and cobble sizes determines the need for crushing, while the amount of fine material determines the need for washing. Petrographic examinations evaluate the nature of aggregate particles and the amount of deleterious material (Meininger and Nichols 1990).

Price and availability are universal criteria that apply to all uses of aggregates. However, the required aggregate characteristics depend on how they will be used in the structure; they may be used as base material, in asphalt concrete, or in portland cement concrete.

4.4 Aggregate Uses

As mentioned, aggregates are primarily used as an underlying material for foundations and pavements and as ingredients in portland cement and asphalt concretes. Aggregate underlying materials, or base courses, can add stability to a structure, provide a drainage layer, and protect the structure from frost damage. Stability is a function of the interparticle friction between the aggregates and the amount of clay and silt "binder" material in the voids between the aggregate particles. However, increasing the clay and silt content will block the drainage paths between the aggregate particles, thereby inhibiting the ability of the material to act as a drainage layer.

In portland cement concrete, 60% to 75% of the volume and 79% to 85% of the weight is made up of aggregates. The aggregates act as a filler to reduce the amount of cement paste needed in the mix. In addition, aggregates have greater volume stability than the cement paste. Therefore, maximizing the amount of aggregate, to a certain extent, improves the quality and economy of the mix.

In asphalt concrete, aggregates constitute over 80% of the volume and 92% to 96% of the mass. The asphalt cement acts as a binder to hold the aggregates together, but does not have enough strength to lock the aggregate particles into position. As a result, the strength and stability of asphalt concrete depends mostly on interparticle friction between the aggregates and, to a limited extent, on the binder.

4.5 Aggregate Properties

Aggregates' properties are defined by the characteristics of both the individual particles and the characteristics of the combined material. These properties can be further described by their physical, chemical, and mechanical characteristics, as shown in Table 4.1 (Meininger and Nichols, 1990). There are several individual particle characteristics that are important in determining if an aggregate source is suitable for a particular application. Other characteristics are measured for designing portland cement and asphalt concrete mixes (Goetz and Wood 1960).

Table 4.1 Basic Aggregate Properties (Meininger and Nichols, 1990)

Property	Relative Importance for End Use*		
	Portland Cement Concrete	Asphalt Concrete	Base
PHYSICAL			
Particle shape (angularity)	M	V	V
Particle shape (flakiness, elongation)	M	M	M
Particle size—maximum	M	M	M
Particle size—distribution	M	M	M
Particle surface texture	M	V	V
Pore structure, porosity	V	M	U
Specific gravity, absorption	V	M	M
Soundness—weatherability	V	M	M
Unit weight, voids—loose, compacted	V	M	M
Volumetric stability—thermal	M	U	U
Volumetric stability—wet/dry	M	U	M
Volumetric stability—freeze/thaw	V	M	M
Integrity during heating	U	M	U
Deleterious constituents	V	M	M
CHEMICAL			
Solubility	M	U	U
Surface charge	U	V	U
Asphalt affinity	U	V	M
Reactivity to chemicals	V	U	U
Volume stability—chemical	V	M	U
Coatings	M	M	U

$$\text{Effective Sp. Gr.} = \frac{\text{Dry weight}}{(\text{Volume not accessible to asphalt})\gamma_w} = \frac{W_s}{(V_s+V_c)\gamma_w} \quad (4.4)$$

where V_c is volume of voids not filled with asphalt cement.

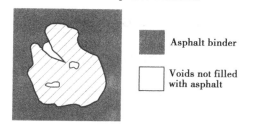

Figure 4.6 Aggregate particle submerged in asphalt cement; not all voids are filled with asphalt.

At present, there is no standard method for determining the effective specific gravity of aggregates directly. The U. S. Corps of Engineers has defined a method for determining the effective specific gravity of aggregates that absorb more than 2.5% water.

The specific gravity and absorption of coarse aggregates are determined in accordance with ASTM C127. In this procedure, a representative sample of the aggregate is soaked for 24 hours and weighed suspended in water. The sample is then dried to the SSD condition and weighed. Finally, the sample is dried to a constant weight and weighed. The specific gravity and absorption are determined by

$$\text{Bulk Dry Sp. Gr.} = \frac{A}{B-C} \quad (4.5)$$

$$\text{Bulk SSD Sp. Gr.} = \frac{B}{B-C} \quad (4.6)$$

$$\text{Apparent Sp. Gr.} = \frac{A}{A-C} \quad (4.7)$$

$$\text{Absorption}(\%) = \frac{B-A}{A}(100) \quad (4.8)$$

where

A = dry weight

B = SSD weight

C = submerged weight

ASTM C128 defines the procedure for determining the specific gravity and absorption of fine aggregates. A representative sample is soaked in water for 24 hours and dried back to the SSD condition. A 500g sample of the SSD material is placed in a pycnometer, a constant volume flask; water is added to the constant volume mark on the pycnometer and the weight is determined again. The sample is then dried and the weight is determined. The specific gravity and absorption are determined by

$$\text{Bulk Dry Sp. Gr.} = \frac{A}{B+S-C} \quad (4.9)$$

$$\text{Bulk SSD Sp. Gr.} = \frac{S}{B+S-C} \quad (4.10)$$

$$\text{Apparent Sp. Gr.} = \frac{A}{B+A-C} \quad (4.11)$$

$$\text{Absorption}(\%) = \frac{S-A}{A}(100) \quad (4.12)$$

where
A = dry weight
B = weight of the pycnometer filled with water
C = weight of the pycnometer filled with aggregate and water
S = saturated surface—dry weight of the sample

4.5.6 Bulk Unit Weight and Voids in Aggregate

The bulk unit weight of aggregate is needed for the proportioning of portland cement concrete mixtures. According to ASTM C29 procedure, a rigid container of known volume is filled with aggregate, which is compacted either by rodding, jigging, or shoveling. The bulk unit weight of aggregate (γ_b) is determined as

$$\gamma_b = \frac{W_s}{V} \quad (4.13)$$

where W_s is the weight of aggregate(stone) and V is the volume of the container.

If the bulk dry specific gravity of the aggregate(G_{sb})(ASTM C127 or C128) is known, the percentage of voids between aggregate particles can be determined as follows:

$$\%V_s = \frac{V_s}{V} \times 100 = \frac{W/\gamma_s}{W/\gamma_b} \times 100 = \frac{\gamma_b}{\gamma_s} \times 100 = \frac{\gamma_b}{G_{sb} \cdot \gamma_w} \times 100$$

$$\%\text{Voids} = 100 - \%V_s \quad (4.14)$$

where
V_s = volume of aggregate
γ_s = unit weight of aggregate
γ_b = bulk unit weight of aggregate
γ_w = unit weight of water

Sample Problem 4.2

Coarse aggregate is placed in a rigid bucket and rodded with a tamping rod to determine its unit weight. The following data are obtained:

Volume of bucket = 1/3 ft³
Weight of empty bucket = 18.5 lb
Weight of bucket filled with dry rodded coarse aggregate = 55.9 lb

a. Calculate the dry-rodded unit weight
b. If the bulk dry specific gravity of the aggregate is 2.630, calculate the percent voids in the aggregate.

Solution

a. Dry-rodded unit weight = (55.9 − 18.5)/0.333 = 112.3 lb/ft³

b. Percent volume of particles = $\dfrac{112.3}{2.630 \times 62.3} \times 100\% = 68.5\%$

Percent voids = 100% − 68.5% = 31.5%

4.5.7 Strength and Modulus

The strength of portland cement concrete and asphalt concrete cannot exceed that of the aggregates. It is difficult and rare to test the strength of aggregate particles. However, tests on the parent rock sample or a bulk aggregate sample provide an indirect estimate of these values. Aggregate strength is generally important in high-strength concrete and in the surface course on heavily traveled pavements. The tensile strength of aggregates ranges from 0.7 MPa to 16 MPa, while the compressive strength ranges from 35 MPa to 350 MPa (Meininger and Nichols 1990; Barksdale 1991). Field service records are a good indication of the adequacy of the aggregate strength.

The modulus of elasticity of aggregates is not usually measured. However, new mechanistic-based methods of pavement design require an estimate of the modulus of aggregate bases. The response of bulk aggregates to stresses is nonlinear and depends on the confining pressure on the material. Since the modulus is used for pavement design, dynamic loads are used in a test to simulate the magnitude and duration of stresses in a pavement base caused by a moving truck. During the test, as the stresses are applied to the sample, the deformation response has two components, a recoverable or resilient deformation, and a permanent deformation. Only the resilient portion of the strain is used with the applied stress level to compute the modulus of the aggregate. Hence, the results are defined as the resilient modulus M_R.

In the resilient modulus test (AASHTO T292), a prepared cylindrical sample is placed in a triaxial cell, as shown in Figure 4.7. A specimen with large aggregates is typically 0.15 m in diameter by 0.30 m high, while soil samples are 71 mm in diameter by 142 mm high. The specimen is subjected to a specified confining pressure and a repeated axial load. Accurate transducers, such as LVDTs, measure the axial deformation. The test requires a determination of the modulus over a range of axial loads and confining pressures. The resilient modulus equals the repeated axial stress divided by the resilient strain for each combination of load level and confining pressure. The resilient modulus test requires the measurement of very small loads and deformations and is, therefore, difficult to perform. Currently, the test is mostly limited to research projects.

4.5.8 Gradation and Maximum Size

Gradation describes the particle size distribution of the aggregate. The particle size distribution is an important attribute of the aggregates. Large aggregates are economically advantageous in portland cement and asphalt concrete, as they have less surface area and, therefore, require less binder. However, large aggregate mixes, whether asphalt or portland cement concrete, are harsher and more difficult to work into place. Hence, construction considerations, such as equipment capability, dimensions of construction members, clearance between reinforcing steel, and layer thickness, limit the maximum aggregate size.

Two definitions are used to describe the maximum particle size in an aggregate blend:

Maximum aggregate size—the smallest sieve size through which 100% of the aggregates sample particles pass.

Nominal maximum aggregate size—the largest sieve that retains any of the aggregate particles, but generally not more than 10%.

Some agencies define the maximum aggregate size as two sizes larger than the first sieve to retain more than 10% of the material, while the nominal maximum size is one size larger than the first sieve to retain more than 10% of the material (The Asphalt Institute 1995; McGennis et al. 1995).

Sieve Analysis Gradation is evaluated by passing the aggregates through a series of sieves, as shown in Figure 4.8 (ASTM C136, E11). The sieve retains particles larger than the opening, while smaller ones pass through. Metric sieve descriptions are based on the size of the openings measured in millimeters. Sieves smaller than 0.6 mm can be described in either millimeters or micrometers. In U.S. customary units, sieves with openings greater than 1/4 in. are designated by the size of the opening; the lengths of the sides of the square openings of a 2-in. sieve are 2

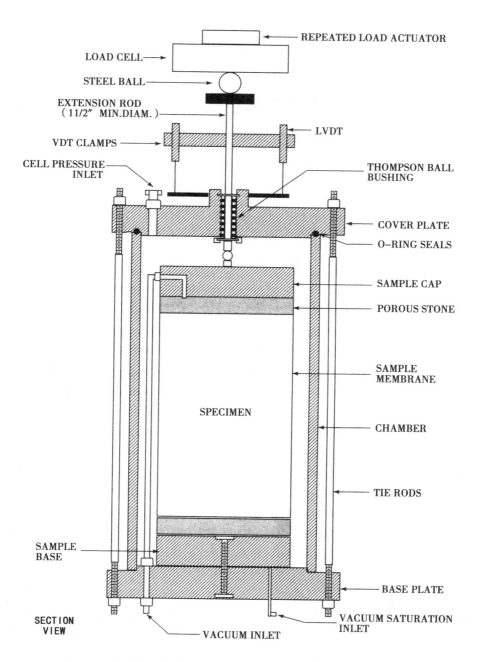

Figure 4.7 Triaxial chamber with external LVDT's and load cell.

Figure 4.8 Sieve shaker for large samples of aggregates.

in. measured between the wires. This equals the diameter of a sphere that will exactly touch each side of the square at the mid-points. Sieves smaller than 1/4 in. are specified by the number of uniform openings per linear inch (a No. 8 sieve has 8 openings per inch, or 64 holes per square inch).

Gradation results are described by the cumulative percentage of aggregates that either pass through or are retained by a specific sieve size. Percentages are reported to the nearest whole number, except that if the percentage passing the 0.075-mm (No. 200) sieve is less than 10%, it is reported to the nearest 0.1%. Gradation analysis results are generally plotted on a semilog chart, as shown in Figures 4.9.

Aggregates are usually classified by size as coarse aggregates, fine aggregates, and mineral fillers (fines). ASTM defines coarse aggregate as particles retained on the 4.75-mm (No. 4) sieve, fine aggregate as those passing the 4.75-mm sieve, and mineral filler as material mostly passing the 0.075-mm (No. 200) sieve.

Maximum Density Gradation The density of an aggregate mix is a function of the size distribution of the aggregates. In 1907 Fuller established the relationship for determining the distribution of aggregates that provides the maximum density or minimum amount of voids as

$$P_i = 100 \left(\frac{d_i}{D}\right)^n \tag{4.15}$$

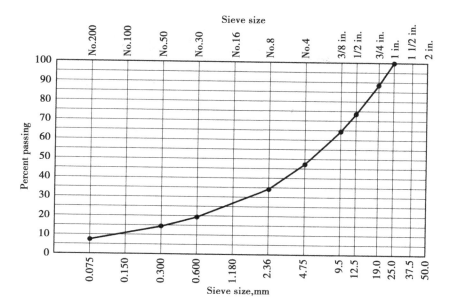

Figure 4.9 Semi-log aggregate gradation chart showing a gradation example. See Table 4.2.

where

P_i = percent passing a sieve of size d_i

d_i = the sieve size in question

D = maximum size of the aggregate

The value of the exponent n recommended by Fuller is 0.5. In the 1960s, the Federal Highway Administration recommended a value of 0.45 for n and introduced the "0.45 power" gradation chart, Figures 4.10, designed to produce a straight line for maximum density gradations(Federal Highway Administration 1988). Table 4.2 presents a sample calculation of the particle size distribution required for maximum density. Note that the gradation in Table 4.2 is plotted on both gradation charts in Figures 4.9 and 4.10.

Table 4.2 Sample Calculations of Aggreagte Distribution Required to Achieve Maximum Density

Sieve	$P_i = 100(d_i/D)^{0.45}$
25 mm(1 in.)	100
19 mm(3/4 in.)	88
12.5 mm(1/2 in.)	73
9.5 mm(3/8 in.)	64
4.75 mm(No. 4)	47

Continued

Sieve	$P_i = 100(d_i/D)^{0.45}$
2.36 mm (No. 8)	34
0.60 mm (No. 30)	19
0.30 mm (No. 50)	14
0.075 mm (No. 200)	7.3

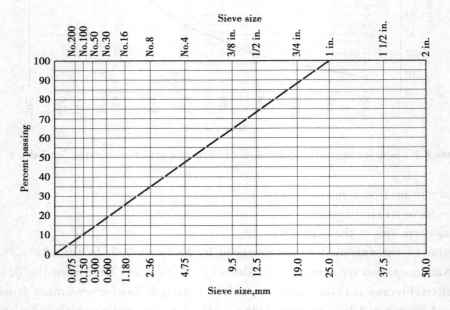

Figure 4.10 Federal Highway Administration 0.45 power gradation chart showing the maximum density gradation for a maximum saze of 25 mm. See Table 4.2.

Frequently, a dense gradation, but not necessarily the maximum possible density, is desired in many construction applications, because of its high stability. Using a high-density gradation also means the aggregates occupy most of the volume of the material, limiting the binder content and thus reducing the cost. For example, aggregates for asphalt concrete must be dense, but must also have sufficient voids in the mineral aggregate to provide room for the binder, plus room for voids in the mixture.

Sample Problem 4.3

A sieve analysis test was performed on a sample of fine aggregate and produced the following results:

Sieve, mm	4.75	2.36	2.00	1.18	0.60	0.30	0.15	0.075	pan
Amount retained, g	0	33.2	56.9	83.1	151.4	40.4	72.0	58.3	15.6

Calculate the percent passing each sieve, and draw a 0.45 power gradation chart with the use of a spreadsheet program.

Solution

Sieve size	Amount Retained, g (a)	Cumulative Amount Retained, g(b)	Cumulative Percent Retained (c)=(b)×100/Total	Percent Passing* (d)=100−(c)
4.75 mm(No. 4)	0	0	0	100
2.36 mm(No. 8)	33.2	33.2	6	94
2.00 mm(No. 10)	56.9	90.1	18	82
1.18 mm(No. 16)	83.1	173.2	34	66
0.60 mm(No. 30)	151.4	324.6	64	36
0.30 mm(No. 50)	40.4	365.0	71	29
0.15 mm(No. 100)	72.0	437.0	86	14
0.075 mm(No. 200)	58.3	495.3	96.9	3.1
Pan	15.6	510.9	100	
Total	510.9			

* Percent passing is computed to a whole percent, except for the 0.075 mm(No. 200) material, which is computed to 0.1%.

The first step in drawing the graph is to compute the sieve size to the 0.45 power, using the metric sieve sizes:

Sieve Size(mm)	Sieve to the 0.45 power	Percent Passing
4.75	2.02	100
2.36	1.47	94
2	1.37	82
1.18	1.08	66
0.6	0.79	36
0.3	0.58	29
0.15	0.43	14
0.075	0.31	3.1

Then the x-y scatter graph function is used to plot the percent passing on the y axis versus the sieve size to the 0.45 power:

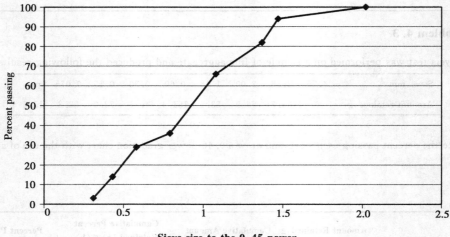

Since the sieve size raised to the 0.45 power is not a meaningful number, the values on the axis are deleted and the text box feature is used to label the x-axis with the actual sieve values. In addition, the drawing tool is used to add vertical lines between the axis and the data points. The resulting graph is as follows:

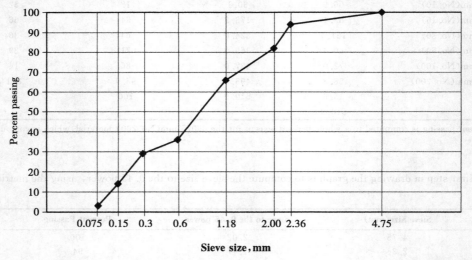

Other Types of Gradation In addition to maximum density (i.e., well-graded), aggregates can have other characteristic distributions, as shown in Figure 4.11. A one-sized distribution has the majority of aggregates passing one sieve and being retained on the next smaller sieve. Hence, the majority of the aggregates have essentially the same diameter; their gradation curve is nearly vertical. One-sized graded aggregates will have good permeability, but poor stability, and are used in such applications as chip seals of pavements. Gapgraded aggregates are missing one or more sizes of material. Their gradation curve has a near horizontal section indicating that nearly the same portions of the aggregates pass two different sieve sizes. Open-graded aggregates are missing small aggregate sizes that would block the voids between the larger aggregate. Since there are a lot of voids, the material will be highly permeable, but may not have good stability.

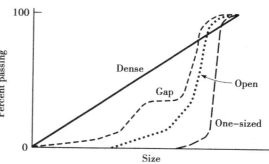

As shown in Table 4.3, the amount of fines has a major effect on the characteristics of aggregate base materials. Aggregates with the percentage of fines equal to the amount required for maximum density have excellent stability and density, but may have a problem with permeability, frost susceptibility, handling, and cohesion.

Figure 4.11 Types of aggregate grain size distributions plotted on a 0.45 gradation chat.

Table 4.3 Effect of Amount of Fines on the Relative Properties of Aggregate Base Material

Characteristic	No fines (Open of Clean)	Well-Graded (Dense)	Large Amount of Fines (Dirty of Rich)
Stability	Medium	Excellent	Poor
Density	Low	High	Low
Permeability	Permeable	Low	Impervious
Frost Susceptibility	No	Maybe	Yes
Handing	Difficult	Medium	Easy
Cohesion	Poor	Medium	Large

Gradation Specifications Gradation specifications define maximum and minimum cumulative percentages of material passing each sieve. Aggregates are commonly described as being either coarse or fine, depending on whether the material is predominantly retained on or passes through a 4.75-mm (No. 4) sieve.

Portland cement concrete requires separate specifications for coarse and fine aggregates. The ASTM C33 specifications for fine aggregates for concrete are given in Table 4.4. Table 4.5 shows the ASTM C33 gradation specifications for coarse concrete aggregates.

Table 4.4 ASTM Gradation Specifications for Fine Aggregates for Portland Cement Concrete (Copyright ASTM, reprinted with permission)

Sieve	Percent Passing
9.5 mm (3/8″)	100
4.75 mm (No. 4)	95~100
2.36 mm (No. 8)	80~100
1.18 mm (No. 16)	50~85
0.60 mm (No. 30)	25~60
0.30 mm (No. 50)	10~30
0.15 mm (No. 100)	2~10

Table 4.5 Coarse Aggregate Grading Requirements for Concrete (ASTM C-33) (Copyright ASTM, reprinted with permission)

Size No.	Nominal Size	Amounts Finer Than Each Laboratory Sieve (Square Openings), Weight Percent												
		4 in. (100 mm)	3 1/2 in. (90 mm)	3 in. (75 mm)	2 1/2 in. (63 mm)	2 in. (50 mm)	1 1/2 in. (37.5 mm)	1 in. (25.0 mm)	3/4 in. (19.0 mm)	1/2 in. (12.5 mm)	3/8 in. (9.5 mm)	No. 4 (4.75 mm)	No. 8 (2.36 mm)	No. 16 (1.18 mm)
1	3 1/2 to 1 1/2 in. (90 to 37.5 mm)	100	90 to 100	...	25 to 60	...	0 to 15	...	...	0 to 5	...	...	...	...
2	2 1/2 to 1 1/2 in. (63 to 37.5 mm)	...	...	100	90 to 100	35 to 70	0 to 15	...	0 to 5	...	...	...	...	...
3	2 to 1 in. (50 to 25.0 mm)	...	...	...	100	90 to 100	35 to 70	0 to 15	...	0 to 5	...	...	...	...
357	2 in. to No. 4 (50 to 4.75 mm)	...	...	...	100	95 to 100	...	35 to 70	...	10 to 30	...	0 to 5	...	...
4	1 1/2 in. to 3/4 in. (37.5 to 19 mm)	...	...	...	...	100	90 to 100	20 to 55	0 to 15	...	0 to 5	...	...	...
467	1 1/2 in. to No. 4 (37.5 to 4.75 mm)	...	...	...	...	100	95 to 100	...	35 to 70	...	10 to 30	0 to 5	...	...

Size No.	Nominal Size											
5	1 to 1/2 in. (25.0 to 12.5 mm)	...	...	100	90 to 100	20 to 55	0 to 10	0 to 5	...	...	...	...
56	1 to 3/8 in. (25.0 to 9.5 mm)	...	...	100	90 to 100	40 to 85	10 to 40	0 to 15	0 to 5	...	...	...
57	1 in. to No. 4 (25.0 to 4.75 mm)	...	...	100	95 to 100	...	25 to 60	...	0 to 10	0 to 5	...	...
6	3/4 in. to 3/8 in. (19.0 to 9.5 mm)	...	...	...	100	90 to 100	20 to 55	0 to 15	0 to 5	...	...	...
67	3/4 in. to No. 4 (19.0 to 4.75 mm)	...	...	...	100	90 to 100	...	20 to 55	0 to 10	0 to 5	...	...
7	1/2 in. to No. 4 (12.5 to 4.75 mm)	...	...	...	...	100	90 to 100	40 to 70	0 to 15	0 to 5	...	...
8	3/8 in. to No. 8 (9.5 to 2.36 mm)	...	...	...	...	...	100	85 to 100	10 to 30	0 to 10	0 to 5	...

Table 4.6 Aggregate Grading Requirements for Superpave Hot Mix Asphalt (AASHTO MP-2)

Sieve Size, mm (in.)	Nominal Maximum Size (mm)					
	37.5	25	19	12.5	9.5	4.75
50 (2 in.)	100	—	—	—	—	—
37.5 (1 1/2 in.)	90—100	100	—	—	—	—
25 (1 in.)	90 max	90—100	100	—	—	—
19 (3/4 in.)	—	90 max	90—100	100	—	—
12.5 (1/2 in.)	—	—	90 max	90—100	100	100
9.5 (3/8 in.)	—	—	—	90 max	90—100	95—100
4.75 (No. 4)	—	—	—	—	90 max	90—100
2.36 (No. 8)	15—41	19—45	23—49	28—58	32—67	—
1.18 (No. 16)	—	—	—	—	—	30—60
0.075 (No. 200)	0.0—6.0	1.0—7.0	2.0—8.0	2.0—10.0	2.0—10.0	6.0—12.0

Generally, local agencies develop their own specifications for the gradation of aggregates for asphalt concrete. Table 4.6 gives the aggregate grading requirements for Superpave hot mix asphalt (McGennis et al. 1995). These specifications define the range of allowable gradations for asphalt concrete for mix design purposes. Note that the percentage of material passing the 0.075-mm (No. 200) sieve, the fines or mineral filler, is carefully controlled for asphalt concrete due to its significance to the properties of the mix.

Once aggregate gradation from asphalt concrete mix design is established for a project, the contractor must produce aggregates that fall within a narrow band around the single gradation line established for developing the mix design. For example, the Arizona Department of Transportation will give the contractor full pay only if the gradation of the aggregates is within the following limits with respect to the accepted mix design gradations:

Sieve Size	Allowable Deviations for Full Pay
9.5 mm (3/8 in.) and larger	±3%
2.36 to 0.45 min (No. 8 to No. 40)	±2%
0.075 mm (No. 200)	±0.5%

Table 4.7 Sample Calculation of Fineness Modulus

Sieve Size	Percentage of Individual Fraction Retained, by Weight	Cumulative Percentage Retained by Weight	Percentage Passing by Weight
9.5 mm (3/8 in.)	0	0	100
4.75 mm (No. 4)	2	2	98
2.36 mm (No. 8)	13	15	85

Continued

Sieve Size	Percentage of Individual Fraction Retained, by Weight	Cumulative Percentage Retained by Weight	Percentage Passing by Weight
1.18 mm(No. 16)	25	40	60
0.60 mm(No. 30)	15	55	45
0.30 mm(No. 50)	22	77	23
0.15 mm(No. 100)	20	97	3
pan	3	100	0
Total	100		
	Fineness Modulus=286/100=2.86		

Fineness Modulus The fineness modulus is a measure of the fine aggregates' gradation and is used primarily for portland cement concrete mix design. It can also be used as a daily quality control check in the production of concrete. The fineness modulus is one-hundredth of the sum of the cumulative percentage weight retained on the 0.15-mm, 0.3-mm, 0.6-mm, 1.18-mm, 2.36-mm, 4.75-mm, 9.5-mm, 19.0-mm, 37.5-mm, 75-mm, and 150-mm(No. 100, 50, 30, 16, 8, 4 and 3/8-in., 3/4-in., $1\frac{1}{2}$-in., 3-in., and 6-in.) sieves. When the fineness modulus is determined for fine aggregates, sieves larger than 9.5 mm are not used. The fineness modulus should be in the range of 2.3 to 3.1, with a higher number being a coarser aggregate. Table 4.7 demonstrates the calculation of the fineness modulus.

Sample Problem 4.4

Calculate the fineness modulus of the sieve analysis results of sample problem 4.3.

Solution

According to the definition of fineness modulus, sieves 2.00 and 0.075 mm(No. 10 and 200) are not included.

$$\text{Fineness modulus} = \frac{6+34+64+71+86}{100} = 26.1$$

Blending Aggregates to Meet Specifications Generally, a single aggregate source is unlikely to meet gradation requirements for portland cement or asphalt concrete mixes. Thus, blending of aggregates from two or more sources would be required to sarisfy the specifications. Figure 4.12 shows a graphical method for selecting the combination of two aggregates to meet a specification. Table 4.8 presents the data used for Figure 4.12. Determining a satisfactory aggregate blend with the graphical method entails the following steps (The Asphalt Institute 1995):

1. Plot the percentages passing through each sieve on the right axis for aggregate A and on the left axis for aggregate B, shown as open circles in Figure 4.12.
2. For each sieve size, connect the left and right axes.
3. Plot the specification limits of each sieve on the corresponding sieve lines; that is, a mark is placed on the 9.5-mm sieve line corresponding to 70% and 90% on the vertical axis, shown as closed circles in Figure 4.12.
4. Connect the upper- and lower-limit points on each sieve line.

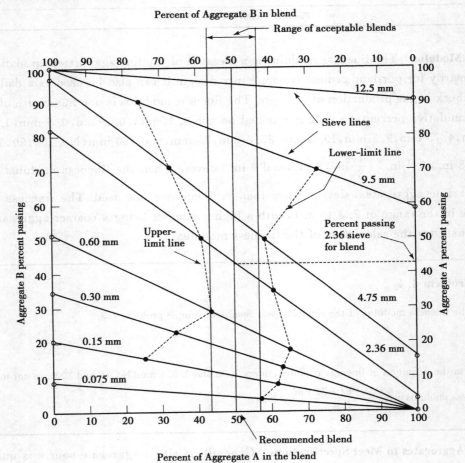

Figure 4.12 Graphical method for determining aggregate blend to meet gradation requirements. See Table 4.8.

5. Draw vertical lines through the rightmost point of the upper-limit line and the leftmost point of the lower-limit line. If the upper-and lower-limit lines overlap, no combination of the aggregates will meet specifications.
6. Any vertical line drawn between these two vertical lines identifies an aggregate blend that will meet the specification. The intersection with the upper axis defines the percentage of aggregate B required for the blend. The projection to the lower axis defines the percentage of aggregate A required.
7. Projecting intersections of the blend line and the sieve lines horizontally gives an estimate of the gradation of the blended aggregate. Figure 4.12 shows that a 50-50 blend of aggregates A and B will result in a blend with 43% passing through the 2.36-mm(No. 8)sieve. The gradation of the blend is shown in the last line of Table 4.8.

When more than two aggregates are required, the graphical procedure can be repeated in an iterative manner. However, a trial and error process is generally used to determine the proportions. The basic equation for blending is

$$P_i = Aa + Bb + Cc + \cdots \quad (4.16)$$

where

P_i = percent blend material passing sieve size i

$A, B, C, \cdots$ = percent of aggregates $A, B, C, \cdots$ passing sieve i

$a, b, c, \cdots$ = decimal fractions by weight of aggregates A, B, and C used in the blend. where the total is 1.00

Table 4.8 Example of Aggregate Blending Analysis by Graphical Method

Sieve	19 mm (3/4 in.)	12.5 mm (1/2 in.)	9.5 mm (3/8 in.)	4.75 mm (No. 4)	2.36 mm (No. 8)	0.60 mm (No. 30)	0.30 mm (No. 50)	0.15 mm (No. 100)	0.075 mm (No. 200)
Specification	100	80—100	70—90	50—70	35—50	18—29	13—23	8—16	4—10
Agg. A	100	90	59	16	3	0	0	0	0
Agg. B	100	100	100	96	82	51	36	21	9
Blend	100	95	80	56	43	26	18	11	4.5

Note: Numbers shown are percent passing each sieve

Table 4.9 demonstrates these calculations for two aggregate sources. The table shows the required specification range and the desired (or target) gradation, usually the midpoint of the specification. A trial percentage of each aggregate source is assumed and is multiplied by the percentage passing each sieve. These gradations are added to get the composite percentage passing each sieve for the blend. The gradation of the blend is compared to the specification range to determine if the blend is acceptable. With practice, blends of four aggregates can readily be resolved. These calculations are easily performed by a spreadsheet computer program.

Table 4.9 Example of Aggregate Blending Analysis by Iterative Method

Sieve	12.5 mm (1/2 in.)	9.5 mm (3/8 in.)	4.75 mm (No. 4)	2.00 mm (No. 10)	0.425 mm (No. 40)	0.180 mm (No. 80)	0.075 mm (No. 200)
Specification	100	95—100	75—85	55—70	20—40	10—20	4—8
Target gradation	100	98	77.5	62.5	30	15	6
%Agg. A(A)	100	100	98	90	71	42	19
%Agg. B(B)	100	94	70	49	14	2	1
30%A(a)	30	30	29.4	27	21.3	12.3	5.7
70%B(b)	70	65.8	49	34.3	9.8	1.4	0.7
Blend(P_i)	100	96	78	61	31	14	6.4

Properties of Blended Aggregates When two or more aggregates from different sources are blended, some of the properties of the blend can be calculated from the properties of the individual components. With the exception of specific gravity and density, the properties of the blend are the simple weighted averages of the properties of the components. This relationship can be expressed as

$$X = P_1 X_1 + P_2 X_2 + P_3 X_3 + \cdots \qquad (4.17a)$$

where

X = composite property of the blend

X_1, X_2, X_3 = properties of fractions 1, 2, 3

P_1, P_2, P_3 = decimal fractions by weight of aggregates 1, 2, 3 used in the blend, where the total is 1.00

This equation applies to properties such as angularity, absorption, strength, and modulus.

Sample Problem 4.5

Coarse aggregates from two stockpiles having coarse aggregate angularity (crushed faces) of 40% and 90% were blended at a ratio of 30 : 70 by weight, respectively. What is the percent of crushed faces of the aggregate blend?

Solution

Crushed faces of the blend = (0.3)(40) + (0.7)(90) = 75%

Equation 4.17a is used for properties that apply to the whole aggregate materials in all stockpiles that are blended. However, some properties apply to either coarse aggregate only or fine aggregate only. Therefore, the percentage of coarse or fine aggregate in each stockpile has to be considered. The relationship in this case is expressed as

$$X = \frac{(x_1 P_1 p_1 + x_2 P_2 p_2 + \cdots + x_n P_n p_n)}{(P_1 p_1 + P_2 p_2 + \cdots + P_n p_n)} \quad (4.17b)$$

where,

X = the test value for the aggregate blend
X_i = the test result for stockpile i
P_i = the percent of stockpile i in the blend
p_i = the percent of stockpile i that either passes or is retained on the dividing sieve

---

Sample Problem 4.6

Aggregates from two stockpiles, A and B having coarse aggregate angularity (crushed faces) of 40% and 90% were blended at a ratio of 30:70 by weight, respectively. The percent material passing the 4.75 mm sieve was 25% and 55% for stockpiles A and B respectively. What is the percent of crushed faces of the aggregate blend?

Solution

Crushed faces of the blend=

$$X = \frac{(x_1 P_1 p_1 + x_2 P_2 p_2 + \cdots + x_n P_n p_n)}{(P_1 p_1 + P_2 p_2 + \cdots + P_n p_n)}$$

$$= \frac{(40 \times 30 \times (100-25) + 90 \times 70 \times (100-55))}{(30 \times (100-25) + 70 \times (100-55))} = 69\%$$

Note that the percentage of coarse aggregate in each stockpile was calculated by subtracting the percentage passing the 4.75 mm sieve from 100.

Asphalt concrete mix design requires that the engineer knows the composite specific gravity of all aggregates in the mix. The composite specific gravity of a mix of different aggregates is obtained by the formula

$$G = \frac{1}{\dfrac{P_1}{G_1} + \dfrac{P_2}{G_2} + \dfrac{P_3}{G_3} + \cdots} \quad (4.18)$$

where

G = composite specific gravity
G_1, G_2, G_3 = specific gravities of fractions 1, 2 and 3
P_1, P_2, P_3 = decimal fractions by weight of aggregates 1, 2 and 3 used in the blend, where the total is 1.00

Note that Equation 4.18 is used only to obtain the combined specific gravity and density of the blend. whereas Equation 4.17 is used to obtain other combined properties.

Sample Problem 4.7

Aggregates from three sources having bulk specific gravities of 2.753, 2.649 and 2.689 were blended at a ratio of 70:20:10 by weight, respectively. What is the bulk specific gravity of the aggregate blend?

Solution

$$G = \frac{1}{\frac{0.7}{2.753} + \frac{0.2}{2.649} + \frac{0.1}{2.689}} = 2.725$$

4.5.9 Deleterious Substances in Aggregate

A deleterious substance is any material that adversely affects the quality of portland cement or asphalt concrete made with the aggregate. Table 4.10 identifies the main deleterious substances in aggregates and their effects on portland cement concrete. In asphalt concrete, deleterious substances are clay lumps, soft or friable particles, and coatings. These substances decrease the adhesion between asphalt and aggregate particles.

Table 4.10 Main Deleterious Substances and Their Affects on Portland Cement Concrete

Substance	Harmful Effect
Organic impurities	Delay settling and hardening, may reduce strength gain, may cause deterioration
Minus 0.075 mm (No. 200) materials	Weaken bond, may increase water requirements
Coal, lignite or other low-density materials	Reduce durability, may cause popouts or stains
Clay lumps and friable particles	Popouts, reduce durability and wear resistance
Soft particles	Reduce durability and wear resistance, popouts

4.5.10 Alkali-Aggregate Reactivity

Some aggregates react with portland cement, harming the concrete structure. The most common reaction, particularly in humid and warm climates, is between the active silica constituents of an aggregate and the alkalis in cement (sodium oxide, Na_2O_2, and potassium oxide, K_2O). The alkali-silica reaction results in excessive expansion, cracking, or popouts in concrete. Other constituents in the aggregate, such as carbonates, can also react with the alkali in the cement; however, their reaction is less harmful. The alkali-aggregate reactivity is affected by the amount, type, and particle size of the reactive material, as well as by the soluble alkali and water content of the concrete.

The best way to evaluate the potential for alkali-aggregate reactivity is by reviewing the field service history. For aggregates without field service history, several laboratory tests are available to check the potential alkali-aggregate reactivity. The ASTM C227 test can be used to determine the potentially expansive alkali-aggregate reactivity of cement-aggregate combinations. In this test, a mortar bar is stored under a prescribed temperature and moisture conditions and its expansion is determined. The quick chemical test (ASTM C289) can be used to identify potentially reactive siliceous aggregates. ASTM C586 is used to determine potentially expansive carbonate rock aggregates (alkali-carbonate reactivity).

If alkali-reactive aggregate must be used, the reactivity can be minimized by limiting the alkali content of the cement. The reactivity can also be reduced by keeping the concrete structure as dry as possible. Fly ash, ground granulated blast furnace slag, silica fume, or natural pozzolans can be used to control the alkali-silica reactivity. Lithium-based admixtures have also been used for the same purpose. Finally, replacing about 30% of a reactive sand-gravel aggregate with crushed limestone (limestone sweetening) can minimize the alkali reactivity (Kosmatka et al. 2002).

4.5.11 Affinity for Asphalt

Stripping, or moisture-induced damage, is a separation of the asphalt film from the aggregate through the action of water. reducing the durability of the asphalt concrete and resulting in pavement failure. The mechanisms causing stripping are complex and not fully understood. One important factor is the relative affinity of the aggregate for either water or asphalt. Hydrophilic (water-loving) aggregates, such as silicates. have a greater affinity for water than for asphalt. They are usually acidic in nature and have a negative surface charge. Conversely, hydrophobic (water-repelling) aggregates have a greater affinity for asphalt than for water. These aggregates, such as limestone, are basic in nature and have a positive surfaca charge. Hydrophilic aggregates are more susceptible to stripping than hydrophobic aggregates. Other stripping factors include porosity, absorption, and the existence of coatings and other deleterious substances.

Since stripping is the result of a compatibility problem between the asphalt and the aggregate, tests for stripping potential are performed on the asphalt concrete mix. Early compatibility tests submerged the sample in either room-temperature water (ASTM D1664) or boiling water (ASTM D3625); after a period of time, the technician observed the percentage of particles stripped from the asphalt. More recent procedures subject asphalt concrete to cycles of freeze-thaw conditioning. The strength or modulus of the specimens is measured and compared with the values of unconditioned specimens(ASTM D1075).

4.6 Sampling Aggregates

Aggregates must be handled and stockpiled in such a way as to minimize segregation, degradation, and contamination. If aggregates roll down the slope of the stockpile, the different sizes will segregate, with large stones at the bottom and small ones at the top. Building stockpiles in thin layers circumvents this problem. The drop height should be limited to avoid breakage, especially for large aggregates. Vibration and jiggling on a conveyor belt tends to work fine material downward while coarse particles rise. Segregation can be minimized by moving the material on the belt frequently(up and down, side to side, in and out)or by installing a baffle plate, rubber sleeve, or paddle wheel at the end of the belt to remix coarse and fine particles. Rounded aggregates segregate more than crushed aggregates. Also, large aggregates segregate more readily than smaller aggregates. Therefore, different sizes should be stockpiled and batched separately. Stockpiles should be separated by dividers or placed in bins to avoid mixing and contamination (Meininger and Nichols 1990).

In order for any of the tests described in this chapter to be valid, the sample of material being tested must represent the whole population of materials that is being quantified with the test. This is a particularly difficult problem with aggregates due to potential segregation problems. Samples of aggregates can be collected from any location in the production process. that is, from the stockpile, conveyor belts, or from bins within the mixing machinery(ASTM D75). Usually, the best location for sampling the aggregate is on the conveyor belt that feeds the mixing plant. However, since the aggregate segregates on the belt, the entire width of the belt should be sampled at several locations or times throughout the production process. The samples would then be mixed to represent the entire lot of material.

Sampling from stockpiles must be performed carefully to minimize segregation. Typically, aggregate samples are taken from the top, middle, and bottom of the stockpile and then combined. Before taking the samples, discard the 75 mm to 150 mm material at the surface. A board shoved vertically into the pile just above the sampling point aids in preventing rolling of coarse aggregates during sampling. Samples are collected using a square shovel and are placed in sample bags or containers and labeled.

Sampling tubes 1.8 m long and 30 mm in diameter are used to sample fine aggregate stockpiles. At least five samples should be collected from random locations in the stockpile. These samples are then combined before laboratory testing.

Field sample sizes are governed by the nominal maximum size of aggregate particles (ASTM D75). Larger-sized aggregates require larger samples to minimize segregation errors. Field samples are typically larger than the samples needed for testing. Therefore, field samples must be reduced using sample splitters or by quartering(Figure 4.13)(ASTM C702).

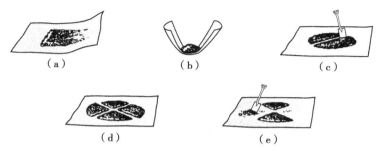

Figure 4.13 Steps for reducing the sample size by quartering: (a) mixing by rolling on blanket, (b) forming a cone after mixing, (c) flattening the cone and quartering, (d) finishing quartering, (e) retaining opposite quarters(the other two quarters are rejected). Copyright ASTM. Reprinted with permission.

SUMMARY

Aggregates are widely used as a base material for foundations and as an ingredient in portland cement concrete and asphalt concrete. While the geological classification of aggregates gives insight into the properties of the material, the suitability of a specific source of aggregates for a particular application requires testing and evaluation. The most significant attributes of aggregates include the gradation, specific gravity, shape and texture, and soundness. When used in concrete, the compatibility of the aggregate and the binder must be evaluated.

QUESTIONS AND PROBLEMS

4.1 What are the three mineralogical or geological classifications of rocks and how are they formed?

4.2 Discuss five different desirable characteristics of aggregate used in portland cement concrete.

4.3 Discuss five different desirable characteristics of aggregate used in asphalt concrete.

4.4 The shape and surface texture of aggregate particles are important for both portland cement concrete and hot mix asphalt.
    a. For preparing PCC, would you prefer round and smooth aggregate or rough and angular aggregate? Briefly explain why(no more than two lines).

b. For preparing HMA, would you prefer round and smooth aggregate or rough and angular aggregate? Briefly explain why (no more than two lines).

4.5 A sample of fine aggregate has the following properties:

Wet mass = 521.0 g

Dry mass = 491.6 g

Absorption = 2.5%

Determine: (a) total moisture content, and (b) free moisture content

4.6 Use the following information to determine the total and free moisture contents in percent:

Mass of wet sand = 627.3 g

Mass of dry sand = 590.1 g

Absorption = 1.5%

4.7 Samples of coarse aggregate from a stockpile are brought to the laboratory for determination of specific gravities. The following weights are found:

Mass of moist aggregate sample as brought to the laboratory: 5298 grams

Mass of oven dried aggregate: 5216 g

Mass of aggregates submerged in water: 3295 g

Mass of SSD (Saturated Surface Dry) Aggregate: 5227 g

Find

a. The aggregate bulk specific gravity

b. The aggregate apparent specific gravity

c. The moisture content of stockpile aggregate (report as a percent)

d. Absorption (report as percent)

4.8 Draw a graph to show the cumulative percent passing through the sieve versus sieve size for well-graded, gap-graded, open-graded, and one-sized aggregates.

4.9 Define the fineness modulus of aggregate. What is it used for?

4.10 What is alkali-silica reactivity? What kind of problems are caused by ASR? Mention two ways to minimize ASR.

4.11 What are the typical deleterious substances in aggregates that affect portland cement concrete? Discuss these effects.

References

Arizona Department of Transportation. *Standard Specifications for Roads and Bridge Construction*. Phoenix, AZ: Arizona Department of Transportation, 2000.

The Asphalt Institute. *Mix Design Methods for Asphalt Concrete and Other Hot-Mix Types*. 6th ed. Manual Series No. 2 (MS-2). Lexington, KY: The Asphalt Institute, 1995.

Barksdale, R. D., ed. *Aggregate Handbook*. Washington, DC: National Stone Association, 1991.

Federal Highway Administration. *Asphalt Concrete Mix Design and Field Control*. Technical Advisory T 5040.27. Washington, DC: Federal High-way Administration, 1988.

Goetz, W. H. and L. E. Wood. Bituminous Materials and Mixtures. *Highway Engineering Handbook*, Section 18. New York: McGraw-Hill, 1960.

Kosmatka, S. H., B. Kerkhoff, and W. C. Panarese. *Design and Control of Concrete Mixtures*. 14th ed. Skokie, IL: Portland Cement Association, 2002.

McGennis, R. B., et al. *Background of Superpave Asphalt Mixture Design and Analysis*. Publication no. FHWA-SA-95-003. Washington, DC: Federal Highway Administration, 1995.

Meininger, R. C. and F. P. Nichols. *Highway Materials Engineering, Aggregates and Unbound Bases*. Publication no. FHWA-HI-90-007. NHI Course No. 13123. Washington, DC: Federal Highway Administration, 1990.

CHAPTER 5

PORTLAND CEMENT

Portland cement concrete is the most widely used manufactured construction material in the world. The importance of concrete in our daily lives cannot be overstated. It is used in structures such as buildings, bridges, tunnels, dams, factories, pavements, and playgrounds. Portland cement concrete consists of portland cement, aggregates, water, air voids, and, in many cases, admixtures. This chapter covers the topics of portland cement, mixing water, and admixtures; Chapter 6 will describe portland cement concrete.

There are many types of concrete, based on different cements. However, portland cement concrete is so prevalent that, unless otherwise identified, the term concrete is always assumed to mean portland cement concrete. Portland cement was patented by Joseph Aspdin in 1824 and was named after the limestone cliffs on the Isle of Portland in England (Kosmatka et al. 2002).

Portland cement is an instant glue (just add water) that bonds aggregates together to make portland cement concrete. Materials specialists concerned with the selection, specification, and quality control of civil engineering projects should understand the production, chemical composition, hydration rates, and physical properties of portland cement.

5.1 Portland Cement Production

Production of portland cement starts with two basic raw ingredients: a calcareous material and an argillaceous material. The calcareous material is a calcium oxide, such as limestone, chalk, or oyster shells. The argillaceous material is a combination of silica and alumina that can be obtained from clay, shale, and blast furnace slag. As shown in Figure 5.1, these materials are crushed and then stored in silos. The raw materials, in the desired proportions, are passed through a grinding mill, using either a wet or dry process. The ground material is stored until it can be sent to the kiln. Modern dry process cement plants use a heat recovery cycle to preheat the ground material, or feed stock, with the exhaust gas from the kiln. In addition, some plants use a flash furnace to further heat the feed stock. Both the preheater and flash furnace improve the energy efficiency of cement production. In the kiln, the raw materials are melted at temperatures of 1400 ℃ to 1650 ℃, changing the raw materials into cement clinker. The clinker is cooled and stored. The final process involves grinding the clinker into a fine powder. During grinding, a small amount of gypsum is added to regulate the setting time of the cement in the concrete.

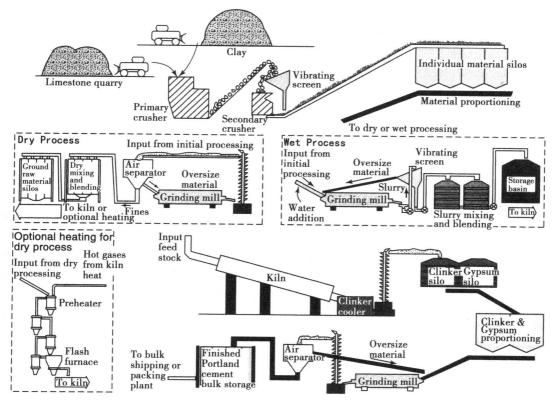

Figure 5.1 Steps in the manufacture of portland cement.

The finished product may be stored and transported in either bulk or sacks. In the United States, a standard sack of cement is 94 lb. which is approximately equal to 1 ft^3 of loose cement when freshly packed. The cement can be stored for long periods of time, provided it is kept dry.

5.2 Chemical Composition of Portland Cement

The raw materials used to manufacture portland cement are lime, silica, alumina, and iron oxide. These raw materials interact in the kiln, forming complex chemical compounds. *Calcination* in the kiln restructures the molecular composition, producing four main compounds, as shown in Table 5.1.

Table 5.1 Main Compounds of Portland Cement.

Compound	Chemical Formula	Common Formula*	Usual Range by Weight(%)
Tricalcium Silicate	$3\,CaO \cdot SiO_2$	C_3S	45~60
Dicalcium Silicate	$2\,CaO \cdot SiO_2$	C_2S	15~30
Tricalcium Aluminate	$3\,CaO \cdot Al_2O_3$	C_3A	6~12
Tetracalcium Aluminoferrite	$4\,CaO \cdot Al_2O_3 \cdot Fe_2O_3$	C_4AF	6~8

* The cement industry commonly uses shorthand notation for chemical formulas: C = Calcium oxide, S = silicon dioxide, A = Aluminum oxide, and F = Iron oxide.

C_3S and C_2S, when hydrated, provide the desired characteristics of the concrete. Alumina and iron, which produce C_3A and C_4AF, are included with the other raw materials to reduce the temperature required to produce C_3S from 2000 ℃ to 1350 ℃. This saves energy and reduces the cost of producing the portland cement.

In addition to these main compounds, there are minor compounds, such as magnesium oxide, titanium oxide, manganese oxide, sodium oxide, and potassium oxide. These minor compounds represent a few percent by weight of cement. The term minor compounds refers to their quantity and not to their importance. In fact, two of the minor compounds, sodium oxide (Na_2O) and potassium oxide (K_2O), are known as alkalis. These alkalis react with some aggregates causing the disintegration of concrete and affecting the rate of strength development, as discussed in Chapter 4.

5.3 Fineness of Portland Cement

Fineness of cement particles is an important property that must be carefully controlled. Since hydration starts at the surface of cement particles, the finer the cement particles, the larger the surface area and the faster the hydration. Therefore, finer material results in faster strength development and a greater initial heat of hydration. Increasing fineness beyond the requirements for a type of cement increases production costs and can be detrimental to the quality of the concrete.

The maximum size of the cement particles is 0.09 mm; 85% to 95% of the particles are smaller than 0.045 mm, and the average diameter is 0.01 mm. (For reference, a number 200 sieve passes material smaller than 0.075 mm.) A kilogram of portland cement has approximately 7 trillion particles with a total surface area of about 300 m^2 to 400 m^2. The total surface area per unit weight is a function of the size of the particles and is more readily measured. Thus, particle size specifications are defined in terms of the surface area per unit weight.

Fineness of cement is usually measured indirectly by measuring the surface area with the Blaine air permeability apparatus (ASTM C204) or the Wagner turbidimeter apparatus (ASTM C115). In the Blaine test, the surface area of the cement particles in cm^2/g is determined by measuring the air permeability of a cement sample and relating it to the air permeability of a standard material. The Wagner turbidimeter determines the surface area by measuring the rate of sedimentation of cement suspended in kerosene. The finer the cement particles, the slower the sedimentation. Both the Blaine and Wagner tests are indirect measurements of surface area and use somewhat different measurement principles. Therefore, tests on a single sample of cement will produce different results. Fineness can also be measured by determining the percent passing the 0.045 mm sieve (No. 325) (ASTM C430).

5.4 Specific Gravity of Portland Cement

The specific gravity of cement is needed for mixture proportioning calculations. The specific gravity of portland cement (without voids between particles) is about 3.15 and can be determined according to ASTM C188. The density of the bulk cement (including voids between particles) varies considerably, depending on how it is handled and stored. For example, vibration during transportation of bulk cement consolidates the cement and increases its bulk density. Thus, cement quantities are specified and measured by weight rather than volume.

5.5 Hydration of Portland Cement

Hydration is the chemical reaction between the cement particles and water. The features of this reaction are the change in matter, the change in energy level, and the rate of reaction. The primary chemical reactions are shown in Table 5.2. Since portland cement is composed of several compounds, many reactions are occurring concurrently.

The hydration process occurs through two mechanisms: through-solution and topochemical. The through-solution process involves the following steps (Mehta and Monteiro 1993):

1. dissolution of anhydrous compounds into constituents
2. formation of hydrates in solution
3. precipitation of hydrates from the supersaturated solution

The through-solution mechanism dominates the early stages of hydration. Topochemical hydration is a solid-state chemical reaction occurring at the surface of the cement particles.

The aluminates hydrate much faster than the silicates. The reaction of tricalcium aluminate with water is immediate and liberates large amounts of heat. Gypsum is used to slow down the rate of aluminate hydration. The gypsum goes into the solution quickly, producing sulfate ions that suppress the solubility of the aluminates. The balance of aluminate to sulfate determines the rate of setting(solidification). Cement paste that sets at a normal rate requires low concentrations of both aluminate and sulfate ions. The cement paste will remain workable for about 45 minutes; thereafter, the paste starts to stiffen as crystals displace the water in the pores. The paste begins to solidify within 2 to 4 hours after the water is added to the cement. If there is an excess of both aluminate and sulfate ions, the workability stare may only last for 10 minutes and setting may occur in 1 to 2 hours. If the availability of aluminate ions is high, and sulfates are low, either a quick set (10 to 45 minutes) or flash set(less than 10 minutes) can occur. Finally, if the aluminate ions availability is low and the sulfate ions availability is high, the gypsum can recrystalize in the pores within 10 minutes, producing a flash set. Flash set is associated with large heat evolution and poor ultimate strength (Mehta and Monteiro 1993).

Calcium silicates combine with water to form calcium-silicate-hydrate, C-S-H. The crystals begin to form a few hours after the water and cement are mixed and can be developed continuously as long as there are unreacted cement particles and free water. C-S-H is not a well-defined compound. The calcium-to-silicate ratio varies between 1.5 and 2.0, and the structurally combined water content is more variable.

As shown in Table 5.2, the silicate hydration produces both C-S-H and calcium hydroxide. Complete hydration of C_3S produces 61% C-S-H and 39% calcium hydroxide; hydration of C_2S results in 82% C-S-H and 18% calcium hydroxide. Since C-S-H is what makes the hydrated cement paste strong, the ultimate strength of the concrete is enhanced by increasing the content of C_2S relative to the amount of C_3S. Furthermore, calcium hydroxide is susceptible to attack by sulfate and acidic waters. Increasing the proportion of C_2S relative to C_3S reduces the quantity of calcium hydroxide and, therefore, improves the durability of the concrete.

Table 5.2 Primary Chemical Reactions During Cement Hydration

$2(3\ CaO \cdot SiO_2)$	+	$6\ H_2O$		=	$3\ CaO \cdot 2\ SiO_2 \cdot 3\ H_2O$ + $3\ Ca(OH)_2$
Tricalcium silicate		Water			Calcium silicate hydrates Calcium hydroxide
$2(2\ CaO \cdot SiO_2)$	+	$4\ H_2O$		=	$3\ CaO \cdot 2\ SiO_2 \cdot 3\ H_2O$ + $Ca(OH)_2$
Dicalcium silicate		Water			Calcium silicate hydrates Calcium hydroxide
$3\ CaO \cdot Al_2O_3$	+	$12\ H_2O$	+ $Ca(OH)_2$	=	$3\ CaO \cdot Al_2O_3 \cdot Ca(OH)_2 \cdot 12\ H_2O$
Tricalcium aluminate		Water	Calcium hydroxide		Calcium aluminate hydrate
$4\ CaO \cdot Al_2O_3 \cdot Fe_2O_3$	+	$10\ H_2O$	+ $2\ Ca(OH)_2$	=	$6\ CaO \cdot Al_2O_3 \cdot Fe_2O_3 \cdot 12\ H_2O$
Tetracalcium aluminoferrite		Water	Calcium Hydroxide		Calcium aluminoferrite hydrate
$3\ CaO \cdot Al_2O_3$	+	$10\ H_2O$	+ $CaSO_4 \cdot 2\ H_2O$	=	$3\ CaO \cdot Al_2O_3 \cdot CaSO_4 \cdot 12\ H_2O$
Tricalcium aluminate		Water	Gypsum		Calcium monosulfoaluminate hydrate

C_3S hydrates more rapidly than C_2S, contributing to the final set time and early strength gain of the cement paste. The rate of hydration is accelerated by sulfate ions in solution. Thus, a secondary effect of the addition of gypsum to cement is to increase the rate of development of the C-S-H.

5.5.1 Structure Development in Cement Paste

The sequential development of the structure in a cement paste is summarized in Figure 5.2. The process begins immediately after water is added to the cement [Figure 5.2(a)]. In less than 10 minutes, the water becomes highly alkaline. As the cement particles hydrate, the volume of the cement particle reduces, increasing the space between the particles. During the early stages of hydration, weak bonds can form, particularly from the hydrated C_3A [Figure 5.2 (b)]. Further hydration stiffens the mix and begins locking the structure of the material in place [Figure 5.2(c)]. Final set occurs when the C-S-H phase has developed a rigid structure, all components of the paste lock into place and the spacing between grains increases as the grains are consumed by hydration [Figure 5.2(d)]. The cement paste continues hardening and gains strength as hydration continues [Figure 5.2(e)]. Hardening develops rapidly at early ages and continues, as long as unhydrated cement particles and free water exist. However, the rate of hardening decreases with time.

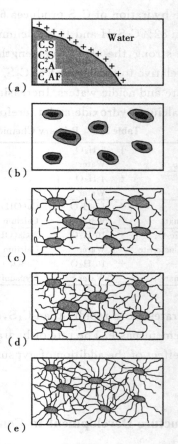

The C-S-H phase is initially formed. C_3A forms a gel fastest.

The volume of cement grain decreases as a gel forms at the surface Cement grains are still able to move independently, but as hydration grows, weak interlocking begins Part of the cement is in a thixotropic state; vibration can break the weak bonds.

The initial set occurs with the development of a weak skeleton in which cement grains are held in place.

Final set occurs as the skeleton becomes rigid, cement particles are locked in place, and spacing between cement grains increases due to the volume reduction of the grains.

Spaces between the cement grains are filled with hydration products as cement paste develops strength and durability.

Figure 5.2 Development of structure in the cement paste: (a) initial C-S-H phase, (b) forming of gels, (c) initial set-development of weak skeleton, (d) final set-development of rigid skeleton, (e) hardening (Hover and Phillco 1990).

5.5.2 Evaluation of Hydration Progress

Several methods are available to evaluate the progress of cement hydration in hardened concrete. These include measuring the following properties (Neville 1981):

1. the heat of hydration
2. the amount of calcium hydroxide in the paste developed due to hydration
3. the specific gravity of the paste
4. the amount of chemically combined water

5. the amount of unhydrated cement paste using X-ray quantitative analysis
6. the strength of the hydrated paste, an indirect measurement

5.6 Voids in Hydrated Cement

Due to the random growth of the crystals and the different types of crystals, voids are left in the paste structure as the cement hydrates. Concrete strength, durability, and volume stability are greatly influenced by voids. Two types of voids are formed during hydration: the interlayer hydration space and capillary voids.

Interlayer hydration space occurs between the layers in the C-S-H. The space thickness is between 0.5 nm and 2.5 nm, which is too small to affect the strength. It can, however, contribute 28% to the porosity of the paste. Water in the interparticle space is strongly held by hydrogen bonds, but can be removed when humidity is less than 11%, resulting in considerable shrinkage.

Capillary voids are the result of the hydrated cement paste having a lower bulk specific gravity than the cement particles. The amount and size of capillary voids depends on the initial separation of the cement particles, which is largely controlled by the ratio of water to cement paste. For a highly hydrated cement paste in which a minimum amount of water was used, the capillary voids will be on the order of 10 nm to 50 nm. A poorly hydrated cement produced with excess water can have capillary voids on the order of 3 mm to 5 mm. Capillary voids greater than 50 nm decrease strength and increase permeability. Removal of water from capillary voids greater than 50 nm does not cause shrinkage, whereas removal of water from the smaller voids causes shrinkage.

In addition to the interlayer space and capillary voids, air can be trapped in the cement paste during mixing. The trapped air reduces strength and increases permeability. However, well-distributed, minute air bubbles can greatly increase the durability of the cement paste. Hence, as described later in this chapter, admixtures are widely used to entrain air into the cement paste.

5.7 Properties of Hydrated Cement

The proper hydration of portland cement is a fundamental quality control issue for cement producers. While specifications control the quality of the portland cement, they do not guarantee the quality of the concrete made with the cement. Mix design, quality control, and the characteristics of the mixing water and aggregates also influence the quality of the concrete. Properties of the hydrated cement are evaluated with either cement paste(water and cement)or mortar(paste and sand).

5.7.1 Setting

Setting refers to the stiffening of the cement paste or the change from a plastic state to a solid state. Although with setting comes some strength, it should be distinguished from hardening, which refers to the strength gain in a set cement paste. Setting is usually described by two levels: initial set and final set. The definitions of the initial and final sets are arbitrary, based on measurements by either the Vicat apparatus(ASTM C191) or the Gillmore needles(ASTM C266).

The Vicat test(Figure 5.3)requires that a sample of cement paste be prepared, using the amount of water required for normal consistency according to a specified procedure. The 1 mm diameter needle is allowed to penetrate the paste for 30 seconds and the amount of penetration is measured. The penetration process is repeated every 15 minutes (every 10 minutes for Type Ⅲ cement)until a penetration of 25 mm or less is obtained. By interpolation, the time when a penetration of 25 mm occurs is determined and recorded as the initial set time. The final set time is when the needle does not penetrate visibly into the paste.

Figure 5.3 Vicat set time apparatus.   **Figure 5.4** Gillmore set time apparatus.

Similar to the Vicat test, the Gillmore test(Figure 5.4)requires that a sample of cement paste of normal consistency be prepared. A pat with a flat top is molded and the initial Gillmore needle is applied lightly to its surface. The application process is repeated until the pat bears the force of the needle without appreciable indentation, and the elapsed time is recorded as the initial set time. This process is then repeated with the final Gillmore needle and the final set time is recorded. Due to the differences in the test apparatuses and procedures, the Vicat and Gillmore tests produce different results for a single sample of material.

The initial set time must allow for handling and placing the concrete before stiffening. The maximum final set time is specified and measured to ensure normal hydration. During cement manufacturing, gypsum is added to regulate the setting time. Other factors that affect the set time include the fineness of the cement, the water-cement ratio, and the use of admixtures.

If the cement is exposed to humidity during storage, a false set might occur in which the cement stiffens within a few minutes of being mixed, without the evolution of much heat. To resolve this problem, the cement paste can be vigorously remixed, without adding water, in order to restore plasticity of the paste and to allow it to set in a normal manner without losing strength. A false set is different than a quick set and a flash set mentioned earlier; a false set can be remedied by remixing, whereas a quick set and a flash set cannot be remedied.

5.7.2 Soundness

Soundness of the cement paste refers to its ability to retain its volume after setting. Expansion after setting, caused by delayed or slow hydration or other reactions, could result if tlle cement is unsound. The autoclave expansion test (Figure 5.5) (ASTM C151) is used to check the soundness of the cement paste. In this test, cement paste bars are subjected to heat and high pressure, and the amount of expansion is measured. ASTM C150 limits antoclave expansion to 0.8%.

Figure 5.5 Cement autoclave expansion apparatus.

5.7.3 Compressive Strength

Compressive strength of mortar is measured by preparing 50-mm cubes and subjecting them to compression according to ASTM C109. The mortar is prepared with cement, water, and standard sand (ASTM C778). Minimum compressive strength values are specified by ASTM C150 for different cement types at different ages. The compressive strength of mortar cubes is proportional to the compressive strength of concrete cylinders. However, the compressive strength of the concrete cannot be predicted accurately from mortar cube strength, since the concrete strength is also affected by the aggregate characteristics, the concrete mixing, and the construction procedures.

5.8 Water-Cementitious Materials Ratio

In 1918, Abrams found that the ratio of the weight of water to the weight of cement, water-cement ratio, influences all the desirable qualities of concrete. For fully compacted concrete made with sound and clean aggregates, strength and other desirable properties are improved by reducing the weight of water used per unit weight of cement. This concept is frequently identified as Abrams's law.

Supplementary cementitious materials, such as fly ash, slag, silica fume, and natural pozzolans, have been used as admixtures in recent years to alter some of the properties of portland cement concrete. Therefore, the term water-cement ratio has been expanded to water-cementitious materials ratio to include these cementitious materials.

Hydration requires approximately 0.22 kg to 0.25 kg of water per 1 kg of cement. Concrete mixes generally require excess moisture, beyond the hydration needs, for workability. Excess water causes the development of capillary voids in the concrete. These voids increase the porosity and permeability of the concrete and reduce strength. Figure 5.6 shows a typical relationship between the water-cementitious materials ratio and compressive strength. It is easy to see that increasing the water-cementitious materials ratio decreases the compressive strength of the concrete for various curing times. A low water-cementitious materials ratio also increases resistance to weathering, provides a good bond between successive concrete layers, provides a good bond between concrete and steel reinforcement, and limits volume change due to wetting and drying. Air-entrained concrete includes an air entraining agent, an admixture, which is used to increase the concrete's resistance to freezing and thawing, as will be discussed later in this chapter. Curing maintains satisfactory moisture content and temperature in the hardened concrete for a definite period of time to allow for hydration (see Chapter 6).

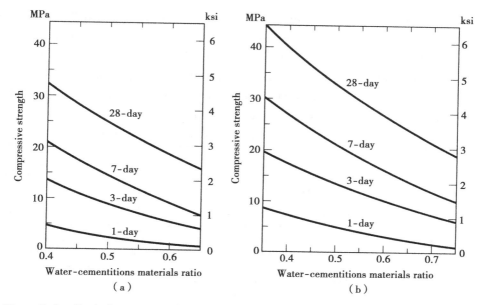

Figure 5.6 Typical age-strength relationships of concrete based on comperssion tests of 0.15×0.30 m (6×12 in.) cylinders, using Type I portland cement and moist-curing at 21C (70F): (a) air entrained concrete, (b) non-air entrained concrete (Kosmatka et al. 1988).

5.9 Types of Portland Cement

Different concrete applications require cements with different properties. some applications require rapid strength gain to expedite the construction process. Other applications require a low heat of hydration to control volume change and associated shrinkage cracking. In some cases, the concrete is exposed to sulfates (SO_4), which can deteriorate normal portland cement concrete. Fortunately, these situations can be accommodated by varying the raw materials used to produce the cement, thereby altering the ratios of the four main compounds of portland cement listed in Table 5.1. The rate of hydration can also be altered by varying the fineness of the cement produced in the final grinding mill. Cement is classified to five standard types, as well as other special types.

5.9.1 Standard Portland Cement Types

Table 5.3 describes the five standard types of portland cement (Types I through V) specified by ASTM C150. In addition to these five types, air entrainers can be added to Type I,

II, and III cements during manufacturing, producing Types I A, II A, and III A, which provide better resistance to freeze and thaw than do non-air-entrained cements. The use of air-entrained cements (Types I A, II A, and III A) has diminished, due to improved availability and reliability of the air entrainer admixtures that can be added during concrete mixing. The uses and effects of air entrainers will be described in the section on admixtures. The ASTM specifications of the standard cement types are shown in Table 5.4.

Table 5.3 Types and Applications of Standard Portland Cement

Type	Name	Application
I	Normal	General concrete work when the special properties of other types are not needed. Suitable for floors, reinforced concrete structures, pavements, etc.
II	Moderate Sulfate Resistance	Protection against moderate sulfate exposure, 0.1%~0.2% weight water soluble sulfate in soil or 150~1500 ppm sulfate in water (sea water). Can be specified with a moderate heat of hydration, making it suitable for large piers, heavy abutments, and retaining walls. The moderate heat of hydration is also beneficial when placing concrete in warm weather.
III	High Early Strength	Used for fast-track construction when forms need to be removed as soon as possible or structure needs to be put in service as soon as possible. In cold weather, reduces time required for controlled curing.
IV	Low Heat of Hydration	Used when mass of structure, such as large dams, requires careful control of the heat of hydration.
V	High Sulfate Resistance	Protection from severe sulfate exposure, 0.2%~2.0% weight water soluble sulfate in soils or 1500~10,800 ppm sulfate in water.

Table 5.4 Standard Properties of Portland Cement (ASTM C150)
(Copyright ASTM, reprinted with permission)

Cement Type	I	I A	II	II A	III	III A	IV	V
Air content of mortar^A, volume %								
max.	12	22	12	22	12	22	12	12
min.	—	16	—	16	—	16	—	—
Fineness^B, specific surface, m^2/kg (alternative methods)								
Turbidimeter test, min.	160	160	160	160	—	—	160	160
Air permeability test, min.	280	280	280	280	—	—	280	280
Autoclave expansion, max., %	0.80	0.80	0.80	0.80	0.80	0.80	0.80	0.80
Minimum compressive strength, psi(MPa)^C								
1 day	—	—	—	—	1800	1450	—	—
	—	—	—	—	(12.4)	(10.0)	—	—
3 days	1800	1450	1500	1200	3500	2800	—	1200

Cement Type	I	I A	II	II A	III	III A	IV	V
	(12.4)	(15.5)	(10.3)	(8.3)	(24.1)	(19.3)	—	(8.3)
			1000^D	800^D				
			(6.9)^D	(5.5)^D				
7 days	2800	2250	2500	2000	—	—	1000	2200
	(19.3)	(15.5)	(17.2)	(13.8)	—	—	(6.9)	(15.2)
			1700^D	1350^D				
			(11.7)^D	(9.3)^D				
28 days	—	—	—	—	—	—	2500	3000
							(17.2)	(20.7)
Time of setting (alternative methods):^{E}								
Gillmore test:								
Initial set, minutes, not less than	60	60	60	60	60	60	60	60
Final set, minutes, not more than	600	600	600	600	600	600	600	600
Vicat test:								
Initial set, minutes, not less than	45	45	45	45	45	45	45	45
Initial set, minutes, not more than	375	375	375	375	375	375	375	375

A. Compliance with the requirements of this specification does not necessarily ensure that the desired air content will be obtained in concrete.
B. Either of the two alternative fineness nethods may be used at the option of the testing laboratory. However, when the sample fails to meet the requirements of the airpermeability test, the turbidimeter test shall be used and the requirements in this table for the turbidimetric method shall govern.
C. The strength at any specified test age shall be not less than that attained at any previous specified test age.
D. When the optional heat of hydration or the chemical limit on the sum of the tricalcium silicate and tricalcium aluminate is specified.
E. The purchaser should specify the type of setting-time test required. In case he does not so specify, the requirements of the Vicat test only shall govern.

The allowable maximum compound compositions are given in Table 5.5, along with the required Blaine fineness (controls particle size). Note that the chemical compositions of Type I and III cements are almost identical; the primary difference is the much greater surface area of the Type III cement. The C_3A contents of Type II and V cements are lower than that of Type I to improve sulfate resistance. C_3S and C_3A are limited in Type IV cement to limit the rate of hydration.

The existence of an ASTM specification for a type of cement does not guarantee that cement's availability. Type I cement is widely available and represents most of the United States' cement production. Type II is the second most available type. Cements can be manufactured that meet all the requirements of both Types I and II; these are labeled Type I/II.

Type Ⅲ cement represents about 4% of U.S. production. Due to the stricter grinding requirements for Type Ⅲ, it is more expensive than Type Ⅰ. The strength gain of Type Ⅰ cement can be accelerated by increasing the cement content per unit volume of concrete, so the selection of Type Ⅲ becomes a question of economics and availability. Type Ⅳ can be manufactured on demand. As discussed later, adding fly ash to Type Ⅰ or Ⅱ portland cement reduces the heat of hydration, producing the benefits of Type Ⅳ, but at a lower cost. Type Ⅴ cement is produced only in locations with a severe sulfate problem.

Table 5.5 ASTM Chemical and Fineness Requirements for Portland Cement (ASTM C150)

Portland Cement Type	Maximum Compound Composition(%)				Blaine Fineness (m^2/kg)
	C_3S	C_2S	C_3A	C_4AF	
Ⅰ	55	19	10	7	370
Ⅱ	51	24	6	11	370
Ⅲ	56	19	10	7	540
Ⅳ	28	49	4	12	380
Ⅴ	38	43	4	9	380

5.9.2 Other Cement Types

Other than the five standard types of portland cement, several hydraulic cements are manufactured in the United States, including

    white portland cement
    blended hydraulic cements
        portland blast furnace slag cement(Type IS)
        portland-pozzolan cement(Type IP and Type P)
        slag cement(Type S)
        pozzolan-modified portland cement(Type I(PM))
        slag-modified portland cement(Type I(SM))
    masonry cements
    expansive cements(Type K)
    specialty cements

    In general, these cements have limited applications. Civil and construction engineers should be aware of their existence, but should study them further before using them.

5.10 Mixing Water

Any potable water is suitable for making concrete. However, some non-potable water may also be suitable. Frequently, material suppliers will use unprocessed surface or well water if it can be obtained at a lower cost than processed water. However, impurities in the mixing water can affect concrete set time, strength, and long-term durability. In addition, chloride ions in the mixing water can accelerate corrosion of reinforcing steel.

5.10.1 Acceptable Criteria

The acceptance criteria for questionable water are specified in ASTM C94. After 7 days, the compressive strength of mortar cubes made with the questionable water should not be less than 90% of the strength of cubes made with potable or distilled water (ASTM C109). Also, the set time of cement paste made with the questionable water should, as measured using the Vicat apparatus (ASTM C191), not be 1 hour less than or 1-1/2 hours more than the set time of paste made with potable or distilled water.

Other adverse effects caused by excessive impurities in mixing water include efflorescence (white stains forming on the concrete surface due to the formation of calcium carbonate), staining, corrosion of reinforcing steel, volume instability, and reduced durability. Therefore, in addition to the compressive strength and set time, there are maximum chemical limits that should not be exceeded in the mixing water, as shown in Table 5.6. Several tests are available to evaluate the chemical impurities of questionable water. Over 100 different compounds and ions can exist in the mixing water and can affect concrete quality; the more important effects are described in the Table 5.7.

Table 5.6 Chemical Limits for Wash Water Used as Mixing Water (ASTM C94)

(Copyright ASTM, reprinted with permission)

Chemical	Maximum Concentration (ppm)	Test Method
Chloride, as Cl		ASTM D512
Prestressed concrete or concrete in bridge decks	500	
Other reinforced concrete in moist environments or containing aluminum embedments or dissimilar metals or with stay-in-place galvanized metal forms	1000	
Sulfate, as SO_4	3000	ASTM D516

Continued

Chemical	Maximum Concentration (ppm)	Test Method
Alkalis, as ($Na_2O + 0.658 K_2O$)	600	
Total Solids	50,000	AASHTO T26

Table 5.7 Summary of Effects of Water Impurities on Concrete Quality

Impurity	Effect
Alkali Carbonate and Bicarbonate	Can retard or accelerate strength test setting and 28-day strength when total dissolved salts exceed 1000 ppm. Can also aggravate alkali-aggregate reaction.
Chloride	Corrosion of reinforcing steel is primary concern. Chloride can enter the mix through admixtures, aggregates, cement, and mixing water, so limits are expressed in terms of total free chloride ions. ACI limits water-soluble ion content based on the type of reinforcement: Prestressed concrete 0.06% Reinforced concrete exposed to chloride in service 0.15% Reinforced concrete protected from moisture 1.00% Other reinforced concrete 0.30%
Sulfate	Can cause expansive reaction and deterioration
Other Salts	Not harmful when concentrations limited to Calcium Bicarbonate 400 ppm Magnesium Bicarbonate 400 ppm Magnesium Sulfate 25,000 ppm Magnesium Chloride 40,000 ppm Iron Salts 40,000 ppm Sodium Sulfide 100 ppm
Sea Water	Do not use for reinforced concrete. Can accelerate strength gain but reduces ultimate strength. Can aggravate alkali reactions.
Acid Water	Limit concentrations of hydrochloric, sulfuric, and other inorganic acids to less than 10,000 ppm.
Alkaline Water	Possible increase in alkali-aggregate reactivity. Sodium hydroxide may introduce quick set at concentrations higher than 0.5%. Strength may be lowered. Potassium hydroxide in concentrations over 1.2% may reduce 28-day strength of some cements.
Sanitary Sewage Water	Dilute to reduce organic matter to less than 20 ppm.
Sugar	Concentrations over 500 ppm can retard setting time and alter strength development. Sucrose in the range of 0.03 to approximately 0.15% usually retards setting. Concentrations over 0.25% by weight of cement can accelerate strength gain, but substantially reduce 28-day strength.
Oils	Mineral oil (petroleum) in excess of 2.5% by weight of mix may reduce strength by 20%.
Algae	Can reduce hydration and entrain air. Do not use water containing algae.

5.10.2 Disposal and Reuse of Concrete Wash Water

Disposal of waste water from ready-mixed concrete operations is a great concern of the ready-mixed concrete producers(Chini and Mbwambo 1996). This waste water is usually generated from truck wash systems, washing of central mixing plant, storm water runoff from the ready-mix plant yard, waste water generated from water sprayed dust control, and conveyor wash down. According to the Water Quality Act(part 116), waste water from ready-mixed concrete operations is a hazardous substance(it contains caustic soda and potash)and its disposal is regulated by the Environmental Protection Agency(EPA). In addition, the high pH makes concrete wash water hazardous under the EPA definition of corrosivity.

The conventional practices for disposing of concrete wash water include dumping at the job site, dumping at a landfill, or dumping into a concrete wash water pit at the ready-mix plant. The availability of landfill sites for the disposal of concrete wash water has been drastically reduced in the past few decades. In addition, the current environmental restrictions either prevent or limit these conventional disposal practices. As a result, most readymix batch plants have developed a variety of operational configurations to manage their own wash water. The alternatives include settling ponds, storm water detention/retention facilities, and water reuse systems. Chemical stabilizing admixture systems have been used to circumvent the necessity to remove any wash water from concrete truck drums, and allow wash water to be reused for mixing more concrete. Studies have concluded that concrete properties are not significantly affected by the use of stabilized wash water(Borger et al. 1994).

Currently, concrete wash water is being used as mixing water for concrete in many places throughout the U. S. Some agencies, however, do not allow its use due to the existence of other impurities derived from concrete admixtures.

5.11 Admixtures for Concrete

Admixtures are ingredients other than portland cement, water, and aggregates that may be added to concrete to impart a specific quality to either the plastic(fresh)mix or the hardened concrete (ASTM C494). Some admixtures are charged into the mix as solutions. In such cases the liquid should be considered part of the mixing water. If admixtures cannot be added in solution, they are either weighed or measured by volume as recommended by the manufacturer. Admixtures are classified by the following chemical and functional physical characteristics (Hewlett 1978):

1. air entrainers
2. water reducers
3. retarders
4. hydration controller admixtures
5. accelerators
6. supplementary cementitious admixtures
7. specialty admixtures

The Portland Cement Association (PCA) identifies four major reasons for using admixtures (Kosmatka et al. 2002):

1. to reduce the cost of concrete construction
2. to achieve certain properties in concrete more effectively than by other means
3. to ensure quality of concrete during the stages of mixing, transporting, placing, and curing in adverse weather conditions
4. to overcome certain emergencies during concrete operations

5.11.1 Air Entrainers

Air entrainers produce tiny air bubbles in the hardened concrete to provide space for water to expand upon freezing. As moisture within the concrete pore structure freezes, three mechanisms contribute to the development of internal stresses in the concrete:

1. Critical saturation—Upon freezing, water expands in volume by 9%. If the percent saturation exceeds 91.7%, the volume increase generates stress in the concrete.
2. Hydraulic pressure—Freezing water draws unfrozen water to it. The unfrozen water moving throughout the concrete pores generates stress, depending on length of flow path, rate of freezing, permeability, and concentration of salt in pores.
3. Osmotic pressure—Water moves from the gel to capillaries to sarisfy thermodynamic equilibrium and to equalize alkali concentrations. Voids permit water to flow from the interlayer hydration space and capillaries into the air voids, where it has room to freeze without damaging the parts.

Internal stresses reduce the durability of hardened concrete, especially when cycles of freeze and thaw are repeated many times. The impact of each of these mechanisms is mitigated by providing a network of tiny air voids in the hardened concrete using air entrainers. In the late 1930s, the introduction of air entrainment in concrete represented a major advance in concrete technology. Air entrainment is recommended for all concrete exposed to freezing.

All concrete contains entrapped air voids, which have diameters of 1 mm or larger and which represent approximately 0.2% to 3% of the concrete volume. Entrained air voids have diameters that range from 0.01 mm to 1 mm, with the majority being less than 0.1 mm. The entrained air voids are not connected and have a total volume between 1% and 7.5% of the concrete volume. Concrete mixed for severe frost conditions should contain approximately 14 billion bubbles per cubic meter. Frost resistance improves with decreasing void size, and small voids reduce strength less than large ones. The fineness of air voids is measured by the specific surface index, equal to the total surfaca area of voids in a unit volume of paste. The specific surface index should exceed 23,600 m^2/m^3 for frost resistance.

In addition to improving durability, air entrainment provides other important benefits to both freshly mixed and hardened concrete. Air entrainment improves concrete's resistance to several destructive factors, including freeze-thaw cycles, deicers and salts, sulfates, and alkali-silica reactivity. Air entrainment also increases the workability of fresh concrete. Air entrainment decreases the strength of concrete, as shown in Figure 5.6, however, this effect can be reduced for moderate-strength concrete by lowering the water-cementitious materials ratio and increasing the cement factor. High strength is difficult to attain with air-entrained concrete.

Air-entraining admixtures are available from several manufacturers and can be composed of a variety of materials, such as

salts of wood resins (Vinsol resin)
synthetic detergents
salts of sulfonated lignin (by-product of paper production)
salts of petroleum acids
salts of proteinaceous material
fatty and resinous acids
alkylbenzene sulfonates
salts of sulfonated hydrocarbons

Air entrainers are usually liquid and should meet the specifications of ASTM C260. The agents enhance air entrainment by lowering the surface tension of the mixing water. Anionic air entrainers are hydrophobic (water hating). The negative charge of the agent is attracted to the positive charge of the cement particle. The hydrophobic agent forms tough, elastic, air-filled bubbles. Mixing disperses the air bubbles throughout the paste, and the sand particles form a grid that holds the air bubbles in place. Other types of air entrainers have different mechanisms but produce similar results.

5.11.2 Water Reducers

Workability of fresh or plastic concrete requires more water than is needed for hydration. The excess water, beyond the hydration requirements, is detrimental to all desirable properties of hardened concrete. Therefore, water-reducing admixtures have been developed to gain workability and, at the same time, maintain quality. Water reducers increase the mobility of the cement particles in the plastic mix, allowing workability to be achieved at lower water contents. Water reducers are produced with different levels of effectiveness: conventional, mid-range, and high-range. Figure 5.7 shows concrete without the addition of admixture and with the addition of conventional, mid-range, and high-range water reducers. As shown in the figure, the slump of the concrete increases, indicating an increase in workability. The highrange water reducer is typically called superplasticizer.

Figure 5.7 Slumps of concretes with the same water-cement ratio: (a) no water reducer, (b) conventional water reducer, (c) mid-range water reducer, (d) highrange water reducer (Superplasticizer).

Water Reducers Mechanism Cement grains develop static electric charge on their surface as a result of the cement-grinding process. Unlike charges attract, causing the cement grains to cluster or "flocculate" [Figure 5.8(a)], which in turn limits the workability. The chemicals in the water-reducing admixtures reduce the static attraction among cement particles. The molecules of water-reducing admixtures have both positive and negative charges at one end, and a single charge (usually negative) on the other end, as illustrated in Figure 5.8(b). These molecules are attracted by the charged surface of the cement grains. The water reducers neutralize the static attraction on the cement surfaces. As a result, the clusters of cement grains are broken apart. Mutual repulsion of like charges pushes the cement grains apart, achieving a better distribution of particles [see Figure 5.8(c)], more uniform hydration, and a less-viscous paste.

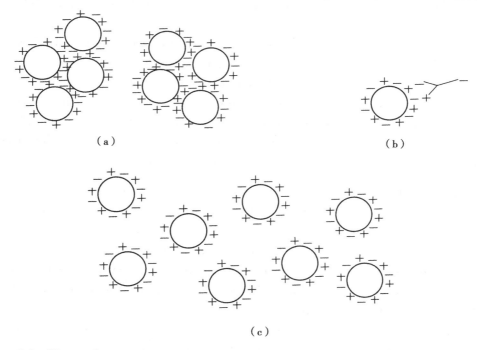

Figure 5.8 Water reducer mechanism: (a) clustering of cement grains without water reducer, (b) molecule of water reducer, and (c) better distribution of cement grains due to the use of water reducer.

Water reducing admixtures can be used indirectly to gain strength. Since the water-reducing admixture increases workability, we can take advantage of this phenomenon to decrease the mixing water, which in turn reduces the water-cementitious materials ratio and increases strength. Hewlett(1978) demonstrated that water reducers can actually be used to accomplish three different objectives, as shown in Table 5.8.

Table 5.8 Effects of Water Reducer

	Cement Content kg/m³	Water-Cement Ratio	Slump(mm)	Compressive Strength(MPa)	
				7 day	28 day
Base Mix	300	0.62	50	25	37
Improve Consistency	300	0.62	100	26	38
Increase Strength	300	0.56	50	34	46
Reduce Cost	270	0.62	50	25.5	37.5

1. Adding a water reducer without altering the other quantities in the mix increases the slump, which is a measure of concrete consistency and an indicator of workability, as will be discussed in Chapter 6.
2. The strength of the mix can be increased by using the water reducer by lowering the quantity of water and keeping the cement content constant.
3. The cost of the mix, which is primarily determined by the amount of cement, can be reduced. In this case, the water reducer allows a decrease in the amount of water. The amount of cement is then reduced to keep the water-cementitious materials ratio equal to the original mix. Thus, the quality of the mix, as measured by compressive strength, is kept constant, although the amount of cement is decreased.

Superplasticizers Superplasticizers, or high-range water reducers, can either greatly increase the flow of the fresh concrete or reduce the amount of water required for a given consistency. For example, adding a superplasticizer to a concrete with a 75-mm slump can increase the slump to 230 mm, or the original slump can be maintained by reducing the water content 12% to 30%. Reducing the amount of mixing water reduces the water-cementitious materials ratio, which in turn, increases the strength of hardened concrete. In fact, the use of superplasticizers has resulted in a major breakthrough in the concrete industry. Now, high-strength concrete in the order of 70~80 MPa compressive strength or more can be produced when superplasticizers are used. Superplasticizers can be used in the following cases:

1. a low water-cementitious materials ratio is beneficial (e.g., high-strength concrete, early strength gain, and reduced porosity)
2. placing thin sections
3. placing concrete around tightly spaced reinforcing steel
4. placing cement underwater
5. placing concrete by pumping
6. consolidating the concrete is difficult

When superplasticizers are used, the fresh concrete stays workable for a short time, 30 min to 60 min, and is followed by rapid loss in workability. Superplasticizers are usually added at the plant to ensure consistency of the concrete. In critical situations, they can be added at the jobsite, but the concrete should be thoroughly mixed following the addition of the admixture. The setting time varies with the type of agents, the amount used, and the interactions with other admixtures used in the concrete.

5.11.3 Retarders

Some construction conditions require that the time between mixing and placing or finishing the concrete be increased. In such cases, retarders can be used to delay the initial set of concrete. Retarders are used for several reasons, such as the following:

1. offsetting the effect of hot weather
2. allowing for unusual placement or long haul distances
3. providing time for special finishes (e.g., exposed aggregate)

Retarders can reduce the strength of concrete at early ages (e.g., one to three days). In addition, some retarders entrain air and improve workability. Other retarders increase the time required for the initial set but reduce the time between the initial and final set. The properties of retarders vary with the materials used in the mix and with job conditions. Thus, the use and effect of retarders must be evaluated experimentally during the mix design process.

5.11.4 Hydration-Control Admixtures

These admixtures have the ability to stop and reactivate the hydration process of concrete. They consist of two parts: a stabilizer and an activator. Adding the stabilizer completely stops the hydration of the cementing materials for up to 72 hours, while adding the activator to the stabilized concrete reestablishes normal hydration and setting. These admixtures are very useful in extending the use of ready-mixed concrete when the work at the jobsite is stopped for various reasons. They are also useful when concrete is being hauled for a long time.

5.11.5 Accelerators

Accelerators are used to develop early strength of concrete at a faster rate than that developed in normal concrete. The ultimate strength, however, of high early strength concrete is about the same as that of normal concrete. Accelerators are used to

1. reduce the amount of time before finishing operations begin
2. reduce curing time
3. increase rate of strength gain
4. plug leaks under hydraulic pressure efficiently

The first three reasons are particularly applicable to concrete work placed during cold temperatures. The increased strength gained helps to protect the concrete from freezing and the rapid rate of hydration generates heat that can reduce the risk of freezing.

Calcium chloride, $CaCl_2$, is the most widely used accelerator (ASTM D98). Both initial and final set times are reduced with calcium chloride. The initial set time of 3 hours for a typical concrete can be reduced to 1.5 hours by adding an amount of calcium chloride equal to 1% of the cement weight; 2% reduces the initial set time to 1 hour. Typical final set times are 6 hours, 3 hours, and 2 hours for 0%, 1%, and 2% calcium chloride. Figure 5.9 shows that strength development is also affected by $CaCl_2$ for plain portland cement concrete (PCC) and portland cement concrete with 2% calcium chloride. Concrete with $CaCl_2$ develops higher early strength compared with plain concrete cured at the same temperature (Hewlett 1978).

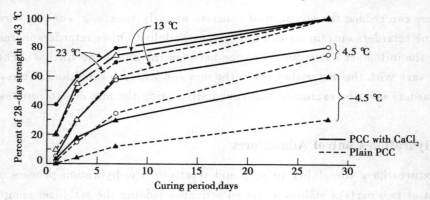

Figure 5.9 Effect of $CaCl_2$ on strength development at different curing temperatures.

The PCA recommends against using calcium chloride under the following conditions:

1. concrete is prestressed
2. concrete contains embedded aluminum such as conduits, especially if the aluminum is in contact with steel
3. concrete is subjected to alkali-aggregate reaction
4. concrete is in contact with water or soils containing sulfates
5. concrete is placed during hot weather
6. mass applications of concrete

The American Concrete Institute (ACI) recommends the following limits to water-soluble chloride ion content based on percent weight of cement (American Concrete Institute 1986):

Member Type	Chloride Ion Limit, %
Prestressed concrete	0.06
Reinforced concrete subjected to chloride in service	0.15
Reinforced concrete protected from moisture	1.00
Other reinforced concrete	0.30

Several alternatives to the use of calcium chloride are available. These include the following:

1. using high early strength (Type III) cement
2. increasing cement content
3. curing at higher temperatures
4. using non-calcium chloride accelerators such as triethanolamine, sodium thiocyanate, calcium formate, calcium nitrite, or calcium nitrate

5.11.6 Supplementary Cementitious Admixtures

Several byproducts of other industries have been used in concrete as supplementary cementitious admixtures since the 1970s, especially in North America. These materials have been used to improve some properties of concrete and to reduce the problem of discarding them. Since these materials are cementitious, they can be used in addition to or as a partial replacement for portland cement. In fact, two or more of these supplementary cementitious additives have been used together to enhance concrete properties. These supplementary cementitious materials include fly ash, ground granulated blast furnace slag, silica fume, and natural pozzolans.

Fly Ash Fly ash is the most commonly used pozzolan in civil engineering structures. Fly ash is a by-product of the coal industry. Combusting pulverized coal in an electric power plant burns off the carbon and most volatile materials. However, depending on the source and type of coal, a significant amount of impurities passes through the combustion chamber. The carbon contents of common coals ranges from 70 to 100 percent. The noncarbon percentages are impurities (e.g., clay, feldspar, quartz, and shale), which fuse as they pass through the combustion chamber. Exhaust gas carries the fused material, fly ash, out of the combustion chamber. The fly ash cools into spheres, which may be solid, hollow (cenospheres), or hollow and filled with other spheres (plerospheres). Particle diameters range from 1 μm to more than 0.1 mm, with an average of 0.015 mm to 0.020 mm, and are 70% to 90% smaller than 0.045 mm. Fly ash is primarily a silica glass composed of silica(SiO_2), alumina(Al_2O_3), iron oxide(Fe_2O_3), and lime (CaO). Fly ash is classified (ASTM C618) as follows:

Class N—Raw or calcined natural pozzolans, including diatomaceous earths, opaline cherts and shales, ruffs and volcanic ashes or pumicites, and some calcined clays and shales

Class F—Fly ash with pozzolan properties

Class C—Fly ash with pozzolan and cementitious properties

Class F fly ash usually has less than 5% CaO but may contain up to 10%. Class C fly ash has 15% to 30% CaO.

The spherical shape of fly ash increases the workability of the fresh concrete. In addition, fly ash extends the hydration process, allowing a greater strength development and reduced porosity. Studies have shown that concrete containing more than 20% fly ash by weight of cement has a much smaller pore size distribution than portland cement concrete without fly ash. The lower heat of hydration reduces the early strength of the concrete. The extended reaction permits a continuous gaining of strength beyond what can be accomplished with plain portland cement.

Ground Granulated Blast Furnace Slag Ground granulated blast furnace slag (GGBF slag) is made from iron blast furnace slag. It is a nonmetallic hydraulic cement consisting basically of silicates and aluminosilicates of calcium, which is developed in a molten condition simultaneously with iron in a blast furnace. The molten slag is rapidly chilled by quenching in water to form a glassy sandlike granulated material. The material is then ground to less than 45 microns. The specific gravity of GGBF slag is in the range of 2.85 to 2.95.

The rough and angular-shaped ground slag in the presence of water and an activator, NaOH or CaOH, both supplied by portland cement, hydrates and sets in a manner similar to portland cement.

Ground slag has been used as a cementitious material in concrete since the beginning of the 1900s. Ground granulated blast furnace slag commonly constitutes between 30% and 45% of the cementing material in the mix. Some slag concretes have a slag component of 70% or more of the cementitious material. ASTM C 989 (AASHTO M 302) classifies slag by its increasing level of reactivity as Grade 80, 100, or 120.

Silica Fume Silica fume is a byproduct of the production of silicon metal or ferrosilicon alloys. One of the most beneficial uses for silica fume is as a mineral admixture in concrete. Because of its chemical and physical properties, it is a very reactive pozzolan. Concrete containing silica fume can have very high strength and can be very durable. Silica fume is available from suppliers of concrete admixtures and, when specified, is simply added during concrete production either in wet or dry forms. Placing, finishing, and curing silica fume concrete require special attention on the part of the concrete contractor.

Silicon metal and alloys are produced in electric furnaces. The raw materials are quartz, coal, and woodchips. The smoke that results from furnace operation is collected and sold as silica fume.

Silica fume consists primarily of amorphous (noncrystalline) silicon dioxide (SiO_2). The individual particles are extremely small, approximately 1/100th the size of an average cement particle. Because of its fine particles, large surfaca area, and the high SiO_2 content, silica fume is a very reactive pozzolan when used in concrete. The quality of silica fume is specified by ASTM C 1240 and AASHTO M 307.

In addition to producing high-strength concrete, silica fume can reduce concrete corrosion induced by deicing or marine salts. Silica fume concrete with a low water content is highly resistant to penetration by chloride ions. More information is available at the www.silicafume.org website.

Natural Pozzolans A pozzolan is a siliceous and aluminous material which, in itself, possesses little or no cementitious value but will, in finely divided form and in the presence of moisture, react chemically with calcium hydroxide at ordinary temperatures to form compounds possessing cementitious properties (ASTM C595). Naturally occurring pozzolans, such as fine volcanic ash, combined with burned lime, were used about 2000 years ago for building construction, and pozzolan continues to be used today. As shown in Table 5.2, calcium hydroxide is one of the products generated by the hydration of C_3S and C_2S. In fact, up to 15% of the weight of portland cement is hydrated lime. Adding a pozzolan to portland cement generates an opportunity to convert this free lime to a cementitious material.

Tables 5.9 and 5.10 summarize the effects of supplementary cementitious admixtures on fresh and hardened concrete. These summaries are based on general trends and should be verified experimentally for specific materials and construction conditions.

Table 5.9 Effect of Supplementary Cementitious Admixtures on Freshly Mixed Concrete

Quality Measure	Effect
Water Requirements	Fly ash reduces water requirements. Silica fume increases water requirements.
Air Content	Fly ash and silica fume reduce air content, compensate by increasing air entrainer.
Workability	Fly ash, ground slag, and inert minerals generally increase workability. Silica fume reduces workability; use superplasticizer to compensate.

138   Materials for Civil and Construction Engineers

Continued

Quality Measure	Effect
Hydration	Fly ash reduces heat of hydration. Silica fume may not affect heat, but superplasticizer used with silica fume can increase heat.
Set Time	Fly ash, natural pozzolans and blast furnace slag increase set time. Can use accelerator to compensate.

Table 5.10 Effect of Supplementary Cementitious Admixtures on Hardened Concrete

Quality Measure	Effect
Strength	Fly ash increases ultimate strength, but reduces rate of strength gain. Silica fume has less effect on rate of strength gain than pozzolans.
Drying shrinkage and creep	Low concentrations usually have little effect. High concentrations of ground slag or fly ash may increase shrinkage. Silica fume may reduce shrinkage.
Permeability and absorption	Generally reduce permeability and absorption. Silica fume is especially effective.
Alkali-aggregate reactivity	Generally reduced reactivity. Extent of improvement depends on type of admixture.
Sulfate resistance	Improved due to reduced permeability.

5.11.7 Specialty Admixtures

In addition to the admixtures previously mentioned, several admixtures are available to improve concrete quality in particular ways. The civil engineer should be aware of these admixtures, but will need to study their application in detail, as well as their cost, before using them. Examples of specialty admixtures include

    workability agents
    corrosion inhibitors
    damp-proofing agents
    permeability-reducing agents
    fungicidal, germicidal, and insecticidal admixtures
    pumping aids
    bonding agents
    grouting agents
    gas-forming agents
    coloring agents
    shrinkage reducing

SUMMARY

The development of portland cement as the binder material for concrete is one of the most important innovations of civil engineering. It is extremely difficult to find civil engineering projects that do not include some component constructed with portland cement concrete. The properties of portland cement are governed by the chemical composition and the fineness of the particles. These control the rate of hydration and the ultimate strength of the concrete. Abrams's discovery of the importance of the water-to-cementitious materials ratio as the factor that controls the quality of concrete is perhaps the single most important advance in concrete technology. Second to this development was the introduction of air entrainment. The subsequent development of additional admixtures for concrete has improved the workability, set time, strength, and economy of concrete construction.

QUESTIONS AND PROBLEMS

5.1 What ingredients are used for the production of portland cement?

5.2 What is the role of gypsum in the production of portland cement?

5.3 What is a typical value for the fineness of portland cement?

5.4 What are the primary chemical reactions during the hydration of portland cement?

5.5 Define the C-S-H phase of cement paste.

5.6 What are the four main chemical compounds in portland cement?

5.7 What chemical compounds contribute to early strength gain?

5.8 Define

   a. interlayer hydration space

   b. capillary voids

   c. entrained air

   d. entrapped air

5.9 Define initial set and final set. Briefly discuss one method used to determine them.

5.10 What is a false set of portland cement? State one reason for false set. If false set is encountered at the job site, what would you do?

5.11 You are an engineer in charge of mixing concrete in an undeveloped area where no potable water is available for mixing concrete. A source of water is available that has some impurities. What tests would you run to evaluate the suitability of this water for concrete mixing? What criteria would you use?

5.12 Discuss the effect of water-to-cementitious materials ratio on the quality of hardened concrete. Explain why this effect happens.

5.13 What type of cement would you use in each of the following cases? Why?

a. Construction of a large pier

b. Construction in cold weather

c. Construction in a warm climate region such as the Phoenix area

d. Concrete structure without any specific exposure condition

e. Building foundation in a soil with severe sulfate exposure

References

American Concrete Institute. *Building Code Requirements for Reinforced Concrete*. ACI Committee 318 Report, ACI 318-99. Farmington Hills, MI: American Concrete Institute, 1999.

Borger, J., R. L. Carrasquillo, and D. W. Fowler. "Use of Recycled Wash Water and Returned Plastic Concrete in the Production of Fresh Concrete." *Advanced Cement Based Materials*, Nov. 1(6), 1994.

Chini, S. A. and W. J. Mbwambo. "Environmentally Friendly Solutions for the Disposal of Concrete Wash Water from Ready Mixed Concrete Operations." *Proceedings, Beijing International Conference*. Beijing, 1996.

Hewlett, P. C. *Concrete Admixtures: Use and Applications*. Rixom, M. R., ed. Lancaster, UK: The Construction Press, 1978.

Hover, K. and R. E. Phillco. *Highway Materials Engineering, Concrete*. Publication No. FHWA-H1-90-009, NHI Course No. 13123. Washington, D.C.: Federal Highway Administration, 1990.

Kosmatka, S. H. and W. C. Panarese. *Design and Control of Concrete Structures*. 13th ed. Skokie, IL: Portland Cement Association, 1988.

Kosmatka, S. H., B. Kerkhoff and W. C. Panarese. *Design and Control of Concrete Mixtures*. 14th ed. Skokie, IL: Portland Cement Association, 2002.

Mehta, P. K. and P. J. M. Monteiro. *Concrete Structure, Properties, and Materials*. 2d ed. Upper Saddle River, NJ: Prentice Hall, 1993.

Neville, A. M. *Properties of Concrete*. 3d ed. London: Pitman Books Ltd, 1981.

CHAPTER 6

PORTLAND CEMENT CONCRETE

Civil and construction engineers are directly responsible for the quality control of portland cement concrete and the proportions of the components used in it. The quality of the concrete is governed by the chemical composition of the portland cement, hydration and development of the microstructure, admixtures, and aggregate characteristics. The quality is strongly affected by placement, consolidation, and curing, as well.

How a concrete structure performs throughout its service life is largely affected by the methods of mixing, transporting, placing, and curing the concrete in the field. In fact, the ingredients of a "good" concrete may be the same as those of a "bad" concrete. The difference, however, is often the expertise of the engineer and technicians who are handling the concrete during construction.

Because of the advances made in concrete technology in the past few decades, concrete can be used in many more applications. Civil and construction engineers should be aware of the alternatives to conventional concrete, such as lightweight concrete, high-strength concrete, polymer concrete, fiber-reinforced concrete, and roller-compacted concrete. Before using these alternatives to conventional concrete, the engineer needs to study them, and their costs, in detail. This chapter covers basic principles of conventional portland cement concrete, its proportioning, mixing and handling, curing, and testing. Alternatives to conventional concrete that increase the applications and improve the performance of concrete are also introduced.

6.1 Proportioning of Concrete Mixes

The properties of concrete depend on the mix proportions and the placing and curing methods. Designers generally specify or assume a certain strength or modulus of elasticity of the concrete when determining structural dimensions. The materials engineer is responsible for assuring that the concrete is properly proportioned, mixed, placed, and cured so to have the properties specified by the designer.

The proportioning of the concrete mix affects its properties in both the plastic and solid states. During the plastic state, the materials engineer is concerned with the workability and finishing characteristics of the concrete. Properties of the hardened concrete important to the materials engineer are the strength, modulus of elasticity, durability, and porosity. Strength is generally the controlling design factor. Unless otherwise specified, concrete strength f'_c refers to the average compressive strength of three tests. Each test is the average result of two 0.15-m×0.30-m cylinders tested in compression after curing for 28 days.

The PCA specifies three qualities required of properly proportioned concrete mixtures (Kosmatka et al. 2002):

1. acceptable workability of freshly mixed concrete
2. durability, strength, and uniform appearance of hardened concrete
3. economy

In order to achieve these characteristics, the materials engineer must determine the proportions of cement, water, fine and coarse aggregates, and the use of admixtures. Several mix design methods have been developed over the years, ranging from an arbitrary volume method (1:2:3 cement:sand:coarse aggregate) to the weight and absolute volume methods prescribed by the American Concrete Institute's Committee 211. The weight method provides relatively simple techniques for estimating mix proportions, using an assumed or known unit weight of concrete. The absolute volume method uses the specific gravity of each ingredient to calculate the unit volume each will occupy in a unit volume of concrete. The absolute volume method is more accurate than the weight method. The mix design process for the weight and absolute volume methods differs only in how the amount of fine aggregates is determined.

6.1.1 Basic Steps for Weight and Absolute Volume Methods

The basic steps required for determining mix design proportions for both weight and absolute volume methods are as follows(Kosmatka et al. 2002):

1. Evaluate strength requirements.
2. Determine the water-cementitious materials ratio required.
3. Evaluate coarse aggregate requirements.
    - ◆ maximum aggregate size of the coarse aggregate
    - ◆ quantity of the coarse aggregate
4. Determine air entrainment requirements.
5. Evaluate workability requirements of the plastic concrete.
6. Estimate the water content requirements of the mix.
7. Determine cementing materials content and type needed.
8. Evaluate the need and application rate of admixtures.
9. Evaluate fine aggregate requirements.
10. Determine moisture corrections.
11. Make and test trial mixes.

Most concrete supply companies have a wealth of experience about how their materials perform in a variety of applications. This experience, accompanied with reliable test data on the relationship between strength and water-cementitious materials ratio, is the most dependable method for selecting mix proportions. However, understanding the basic principles of mixture design and the proper selection of materials and mixture characteristics is as important as the actual calculation. Therefore, the PCA procedure provides guidelines and can be adjusted to match the experience obtained from local conditions. The PCA mix design steps are discussed next.

1. **Strength Requirements** Variations in materials, and batching and mixing of concrete results in deviations in the strength of the concrete produced by a plant. Generally, the structural design engineer does not consider this variability when determining the size of the structural members. If the materials engineer provides a material with an average strength equal to the strength specified by the designer, then half of the concrete will be weaker than the specified strength. Obviously, this is undesirable. To compensate for the variance in concrete strength, the materials engineer designs the concrete to have an average strength greater than the strength specified by the structural engineer.

In order to compute the strength requirements for concrete mix design, three quantities must be known:

1. the specified compressive strength f'_c
2. the variability or standard deviation s of the concrete
3. the allowable risk of making concrete with an unacceptable strength

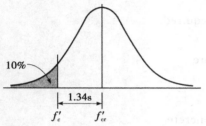

Figure 6.1 Use of normal distribution and risk criteria to estimate average required concrete strength.

The standard deviation in the strength is determined for a plant by making batches of concrete, testing the strength for many samples, and computing the standard deviation using Equation 1.15 in Chapter 1. The allowable risk has been established by the American Concrete Institute (ACI). One of the risk rules states that there should be less than 10% chance that the strength of a concrete mix is less than the specified strength.

Assuming that the concrete strength has a normal distribution, the implication of the ACI rule is that 10% of the area of the distribution must be to tlle left of f'_c, as shown in Figure 6.1. Using a table of standard z values for a normal distribution curve, we can determine that 90% of the area under the curve will be to the right of f'_c if the average strength is 1.34 standard deviations from f'_c. In other words, the required average strength f'_{cr} for this criterion can be calculated as

$$f'_{cr} = f'_c + 1.34s \tag{6.1}$$

where

f'_{cr} = required average compressive strength, MPa or psi
f'_c = specified compressive strength, MPa or psi
s = standard deviation, MPa or psi

For mixes with a large standard deviation in strength, the ACI has another risk criterion that requires

$$f'_{cr} = f'_c + 2.33s - 3.45 \tag{6.2}$$

The required average compressive strength f'_{cr} is determined as the larger value obtained from Equations 6.1 and 6.2.

Equation 6.2 is valid for SI units only. If U.S. customary units are used, f'_{cr}, f'_c, and s are recorded in psi and the constant 3.45 in Equation 6.2 should be changed to 500.

The standard deviation should be determined from at least 30 strength tests. If the standard deviation is computed from 15 to 30 samples, then the standard deviation is multiplied by the following factor, F, to determine the modified standard deviation s'.

Number of Tests	Modification Factor F
15	1.16
20	1.08
25	1.03
30 or more	1.00

Linear interpolation is used for an intermediate number of tests, and s' is used in place of s in Equations 6.1 and 6.2.

If fewer than 15 tests are available, the following adjustments are made to the specified strength, instead of using Equations 6.1 and 6.2:

Specified Compressive Strength f'_c, MPa (psi)	Required Average Compressive Strength f'_{cr}, MPa (psi)
<21 (<3000)	$f'_c + 7.0$ ($f'_c + 1000$)
21 to 35 (3000 to 5000)	$f'_c + 8.5$ ($f'_c + 1200$)
>35 (>5000)	$f'_c + 10.0$ ($f'_c + 1400$)

These estimates are very conservative and should not be used for large projects, since the concrete will be overdesigned and, therefore, not economical.

Sample Problem 6.1

The design engineer specifies a concrete strength of 31.0 MPa. Determine the required average compressive strength for

a. a new plant for which s is unknown
b. a plant for which $s = 3.6$ MPa for 17 test results
c. a plant with extensive history of producing concrete with $s = 2.4$ MPa
d. a plant with extensive history of producing concrete with $s = 3.8$ MPa

Solution

a. $f'_{cr} = f'_c + 8.5 = 31.0 + 8.5 = 39.5$ MPa

b. Need to interpolate modification factor:
$$F = 1.16 - \left(\frac{1.16 - 1.08}{20 - 15}\right)(17 - 15) \cong 1.13$$

Multiply standard deviation by the modification factor
$$s' = (s)(F) = 3.6(1.13) = 4.1 \text{ MPa}$$

Determine maximum from Equations 6.1 and 6.2
$$f'_{cr} = 31.0 + 1.34(4.1) = 36.5 \text{ MPa}$$

$f'_{cr} = 31.0 + 2.33(4.1) - 3.45 = 37.1$ MPa

Use $f'_{cr} = 37.1$ MPa

c. Determine maximum from Equations 6.1 and 6.2

$f'_{cr} = 31.0 + 1.34(2.4) = 34.2$ MPa

$f'_{cr} = 31.0 + 2.33(2.4) - 3.45 = 33.1$ MPa

Use $f'_{cr} = 34.2$ MPa

d. Determine maximum from Equations 6.1 and 6.2

$f'_{cr} = 31.0 + 1.34(3.8) = 36.1$ MPa

$f'_{cr} = 31.0 + 2.33(3.8) - 3.45 = 36.4$ MPa

Use $f'_{cr} = 36.4$ MPa

---

2. Water-Cementitious Materials Ratio Requirements The next step is to determine the water-cementitious materials ratio needed to produce the required strength. Historical records are used to plot a strength-versus-water-cementitious materials ratio curve, such as that seen in Figure 6.2. If historical data are not available, three trial batches are made at different water-cementitious materials ratios to establish a curve similar to Figure 6.2. Table 6.1 can be used for estimating the water-cementitious materials ratios for the trial mixes when no other data are available. The required average compressive strength is used with the strength versus water-cementitious materials relationship to determine the water-cementitious materials ratio required for the strength requirements of the project.

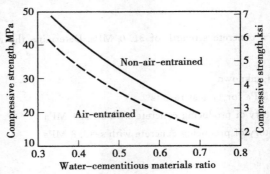

Figure 6.2 Example trial mixture or field data strength curves.

For small projects of noncritical applications, Table 6.2 can be used in lieu of trial mixes, with the permission of the project engineer. Table 6.2 is conservative with respect to the strength versus water-cementitious materials ratio relationship. This results in higher cement factors and greater average strengths than would be required if a mix design is performed. This table is not intended for use in designing trial batches; use Table 6.1 for trial batch design.

Table 6.1 Typical Relationship between Water-Cementitious Materials Ratio and Compressive Strength of Concrete*

Compressive Strength at 28 days, MPa(psi)**	Water-Cementitious Materials Ratio by Weight	
	Non-Air-Entrained Concrete	Air-Entrained Concrete
48(7000)	0.33	—
41(6000)	0.41	0.32
35(5000)	0.48	0.40
28(4000)	0.57	0.48
21(3000)	0.68	0.59
14(2000)	0.82	0.74

* American Concrete Institute (ACI 211.1 and ACI 211.3)
** Strength is based on cylinders moist-cured 28 days in accordance with ASTM C31(AASHTO T23). Relationship assumes nominal maximum size of aggregate about 19 to 25 mm.

Table 6.2 Maximum Permissible Water-Cementitious Materials Ratios for Concrete when Strength Data from Field Experience or Trial Mixtures are not Available*

Specified 28-day compressive, Strength, f'_c, MPa(psi)	Water-Cementitious Materials Ratio by Weight	
	Non-Air-Entrained Concrete	Air-Entrained Concrete
17(2500)	0.67	0.54
21(3000)	0.58	0.46
24(3500)	0.51	0.40
28(4000)	0.44	0.35
31(4500)	0.38	**
35(5000)	**	**

* American Concrete Institute (ACI 318), 1999.
** For strength above 31.0 MPa(4500 psi)(non-air-entrained concrete) and 27.6 MPa(4000 psi)(air-entrained concrete), concrete proportions shall be established from field data or trial mixtures.

The water-cementitious materials ratio required for strength is checked against the maximum allowable water-cementitious materials ratio for the exposure conditions. Tables 6.3 and 6.4 provide guidance on the maximum allowable water-cementitious materials ratio and the minimum design compressive strength for exposure conditions. Generally, more severe exposure conditions require lower water-cementitious materials ratios. The minimum of the water-cementitious materials ratio for strength and exposure is selected for proportioning the concrete.

If a pozzolan is used in the concrete, the water-cementitious materials plus pozzolan ratio by weight may be used instead of the traditional water-cementitious materials ratio. In other words, the weight of the water is divided by the sum of the weights of cement plus pozzolan.

Table 6.3 Maximum Water-Cementitious Material Ratios and Minimum Design Strengths for Various Exposure Conditions*

Exposure Condition	Maximum Water-Cementitious Material Ratio by Mass for Concrete	Minimum Design Compressive Strength, f'_c, MPa(psi)
Concrete protected from exposure to freezing and thawing, application of deicing chemicals, or aggressive substances	Select water-cementitious material ratio on basis of strength, workability, and finishing needs	Select strength based on structural requirements
Concrete intended to have low permeability when exposed to water	0.50	28(4000)
Concrete exposed to freezing and thawing in a moist condition or deicers	0.45	31(4500)
For corrosion protection for reinforced concrete exposed to chlorides from deicing salts, salt water, brackish water, seawater, or spray from these sources	0.40	35(5000)

* American Concrete Institute(ACI 318), 1999.

Table 6.4 Requirements for Concrete Exposed to Sulfates in Soil or Water*

Sulfate Exposure	Water-Soluble Sulfate(SO_4) in Soil, Percent by Weight**	Sulfate (SO_4) in Water, ppm**	Cement Type***	Maximum Water-Cementitious Material Ratio by Weight	Minimum Design Compressive Strength, f'_c, MPa(psi)
Negligible	Less than 0.10	Less than 150	No special type required	—	—
Moderate****	0.10~0.20	150~1500	II, MS, IP(MS), IS(MS), P(MS), I(PM)(MS), I(SM)(MS)	0.50	28(4000)
Severe	0.20~2.00	1500~10,000	V, HS	0.45	31(4500)
Very Severe	Over 2.00	Over 10,000	V, HS	0.40	35(5000)

* Adopted from American Concrete Institute(ACI 318), 1999.
** Tested in accordance with the Method for Determining the Quantity of Soluble Sulfate in Solid (Soil and Rock) and Water Samples, Bureau of Reclamation, Denver, 1977.
*** Cement Types II and V are in ASTM C150(AASHTO M85), Types MS and HS in ASTM C1157, and the remaining types are in ASTM C595(AASHTO M240). Pozzolans or slags that have been determined by test or severe record to improve sulfate resistance may also be used.
**** Sea water.

3. **Coarse Aggregate Requiements** The nest step is to determine the suitable aggregate characteristics for the project. In general, large dense graded aggregates provide the most economical mix. Large aggregates minimize the amount of water required and, therefore, reduce the a-

mount of cement required per cubic meter of mix. Round aggregates require less water than angular aggregates for an equal workability.

The maximum allowable aggregate size is limited by the dimensions of the structure and the capabilities of the construction equipment. The largest maximum aggregate size practical under job conditions that satisfies the size limits in the table shonld be used.

Situation	Maximum Aggregate Size
Form dimensions	1/5 of minimum clear distance
Clear space between reinforcement or prestressing tendons	3/4 of minimum clear space
Clear space between reinforcement and form	3/4 of minimum clear space
Unreinforced slab	1/3 of thickness

Sample Problem 6.2

A structure is to be built with concrete with a minimum dimension of 0.2 m, minimum space between rebars of 40 mm, and minimum cover over rebars of 40 mm. Two types of aggregate are locally available, with maximum sizes of 19 mm and 25 mm, respectively. If both types of aggregate have essentially the same cost, which one is more suitable for this structure?

Solution

25 mm < (1/5)(200 mm) minimum dimensions.
25 mm < (3/4)(40 mm) rebar spacing.
25 mm < (3/4)(40 mm) rebar cover.

Therefore, both sizes satisfy the dimension requirements. However, 25 mm aggregate is more suitable, because it will produce more economical concrete mix.

The gradation of the fine aggregates is defined by the fineness modulus. The desirable fineness modulus depends on the coarse aggregate size and the quantity of cement paste. A low fineness modulus is desired for mixes with low cement content to promote workability.

Once the fineness modulus of the fine aggregate and the maximum size of the coarse aggregate are determined, the volume of coarse aggregate per unit volume of concrete is determined using Table 6.5. For example, if the fineness modulus of the fine aggregate is 2.60 and the maximum aggregate size is 25 mm, the coarse aggregate will have a volume of 0.69 m^3/m^3 of concrete. Table 6.5 is based on the unit weight of aggregates in a dry-rodded condition (ASTM C29). The values given are based on experience in producing an average degree of workability. The volume of coarse aggregate can be increased by 10% when less workability is required, such as in pavement construction. The volume of coarse aggregate should be reduced by 10% to increase workability, for example to allow placement by pumping.

Table 6.5 Bulk Volume of Coarse Aggregate per Unit Volume of Concrete*

Nominal Maximum Size of Aggregate, mm(in.)	Bulk Volume of Dry-Rodded Coarse Aggregate Per Unit Volume of Concrete for Different Fineness Moduli of Fine Aggregate**			
	Fineness Modulus			
	2.40	2.60	2.80	3.00
9.5(3/8)	0.50	0.48	0.46	0.44
12.5(½)	0.59	0.57	0.55	0.53
19(¾)	0.66	0.64	0.62	0.60
25(1)	0.71	0.69	0.67	0.65
37.5(1½)	0.75	0.73	0.71	0.69
50(2)	0.78	0.76	0.74	0.72
75(3)	0.82	0.80	0.78	0.76
150(6)	0.87	0.85	0.83	0.81

* American Concrete Institute (ACI 211.1).
** Bulk volumes are based on aggregates in a dry-rodded condition as described in ASTM C29 (AASHTO T19).

4. Air Entrainment Requirements Next, the need for air entrainment is evaluated. Air entrainment is required whenever concrete is exposed to freeze-thaw conditions and deicing salts. Air entrainment is also used for workability in some situations. The amount of air required varies based on exposure conditions and is affected by the size of the aggregates. The exposure levels are defined as follows:

Mild exposure—Indoor or outdoor service in which concrete is not exposed to freezing and deicing salts. Air entrainment may be used to improve workability.

Moderate exposure—Some freezing exposure occurs, but concrete is not exposed to moisture or free water for long periods prior to freezing. Concrete is not exposed to deicing salts. Examples include exterior beams, columns, walls, etc., not exposed to wet soil.

Severe exposure—Concrete is exposed to deicing salts, saturation, or free water. Examples include pavements, bridge decks, curbs, gutters, canal linings, etc.

Table 6.6 presents the recommended air contents for different combinations of exposure conditions and maximum aggregate sizes. The values shown in Table 6.6 are the entrapped air for non-air-entrained concrete and the entrapped plus entrained air in case of air-entrained concrete. The recommended air content decreases with increasing maximum aggregate size.

Table 6.6 Approximate Target Percent Air Content Requirements for Different Nominal Maximum Sizes of Aggregates*

	Maximum Aggregate Size							
	9.5 mm (3/8 in.)	12.5 mm (½ in.)	19 mm (¾ in.)	25 mm (1 in.)	37.5 mm (1½ in.)	50 mm (2 in.)	75 mm (3 in.)	150 mm (6 in.)
Non-air-entrained concrete	3	2.5	2	1.5	1	0.5	0.3	0.2
Air-entrained concrete**								
Mild Exposure	4.5	4.0	3.5	3.0	2.5	2.0	1.5	1.0
Moderate Exposure	6.0	5.5	5.0	4.5	4.5	4.0	3.5	3.0
Severe Exposure	7.5	7.0	6.0	6.0	5.5	5.0	4.5	4.0

* American Concrete Institute(ACI 211.1 and ACI 318).

** The air content in job specifications should be specified to be delivered within −1 to +2 percentage points of the table target value for moderate and severe exposures.

5. Workability Requirements The next step in the mix design is to determine the workability requirements for the project. Workability is defined as the ease of placing, consolidating, and finishing freshly mixed concrete. Concrete should be workable but should not segregate or excessively bleed(migration of water to the top surface of concrete). The slump test(Figure 6.3) is an indicator of workability when evaluating similar mixtures. This test consists of filling a truncated cone with concrete, removing the cone, then measuring the distance the concrete slumps(ASTM C143). The slump is increased by adding water, air entrainer, water reducer, superplasticizer, or by using round aggregates. Table 6.7 provides recommendations for the slump of concrete used in different types of projects. For batch adjustments, slump increases about 25 mm for each 6 kg of water added per m³ of concrete.

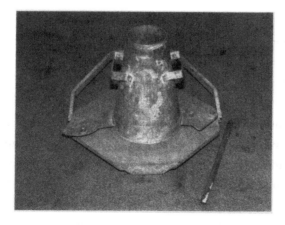

Figure 6.3 Slump test apparatus.

Table 6.7 Recommended Slumps for Various Types of Construction*

Concrete Construction	Slump, mm (In.)	
	Maximum**	Minimum
Reinforced foundation walls and footings	75(3)	25(1)
Plain footings, caissons, and substructure walls	75(3)	25(1)
Beams and reinforced walls	100(4)	25(1)
Building columns	100(4)	25(1)
Pavements and slabs	75(3)	25(1)
Mass concrete	75(3)	25(1)

* American Concrete Institute (ACI 211.1).

** May be increased 25 mm (1 in.) for consolidation by hand methods such as rodding and spading. Plasticizers can safely provide higher slumps.

6. **Water Content Requirements** The water content required for a given slump depends on the maximum size and shape of the aggregates and whether an air entrainer is used. Table 6.8 gives the approximate mixing water requirements for angular coarse aggregates (crushed stone). The recommendations in Table 6.8 are reduced for other aggregate shape as shown in this table.

Aggregate Shape	Reduction in Water Content, kg/m^3 (lb/yd^3)
Subangular	12(20)
Gravel with crushed particles	21(35)
Round gravel	27(45)

These recommendations are approximate and should be verified with trial batches for local materials.

7. **Cementing Materials Content Requirements** With the water-cementitious materials ratio and the required amount of water estimated, the amount of cementing materials required for the mix is determined by dividing the weight of the water by the water-cementitious materials ratio. PCA recommends a minimum cement content of 334 kg/m^3 for concrete exposed to severe freeze-thaw, deicers, and sulfate exposures, and not less than 385 kg/m^3 for concrete placed under water. In addition, Table 6.9 shows the minimum cement requirements for proper placing, finishing, abrasion resistance, and durability in flatwork, such as slabs.

Table 6.8 Approximate Mixing Water in kg/m³ (lb/yd³) for Different Slumps and Nominal Maximum Aggregate Sizes*

Slump, mm(in.)	Maximum Aggregate Size in mm(in.)**							
	9.5(3/8)	12.5(1/2)	19(3/4)	25(1)	37.5(1½)	50(2)***	75(3)***	150(6)***
	Non-air-entrained concrete							
25 to 50 (1 to 2)	207(350)	199(335)	190(315)	179(300)	166(275)	154(260)	130(220)	113(190)
75 to 100 (3 to 4)	228(385)	216(365)	205(340)	193(325)	181(300)	169(285)	145(245)	124(210)
150 to 175 (6 to 7)	243(410)	228(385)	216(360)	202(340)	190(315)	178(300)	160(270)	—
	Air-entrained concrete							
25 to 50 (1 to 2)	181(305)	175(295)	168(280)	160(270)	150(250)	142(240)	122(205)	107(180)
75 to 100 (3 to 4)	202(340)	193(325)	184(305)	175(295)	165(275)	157(265)	133(225)	119(200)
150 to 175 (6 to 7)	216(365)	205(345)	197(325)	184(310)	174(290)	166(280)	154(260)	—

* American Concrete Institute (ACI 211.1 and ACI 318).

** These quantities of mixing water are for use in computing cementitious material contents for trial batches. They are maximums for reasonably well-shaped angular coarse aggregates graded within limits of accepted specifications.

*** The slump values for concrete containing aggregates larger than 37.5 mm(1½ in.) are based on slump tests made after removal of particles larger than 37.5 mm by wet screening.

8. Admixture Requirements If one or more admixtures are used to add a apecific quality in the concrete (as discussed in Chapter 5), their quantities should be considered in the mix proportioning. Admixture manufacturers provide specific information on the quantity of admixture required to achieve the desired results.

9. Fine Aggregate Requirements At this point, water, cement, and coarse aggregate weights per cubic meter(cubic yard) are known and the volume of air is estimated. The only remaining factor is the amount of fine aggregates needed. The weight mix design method uses Table 6.10 to estimate the total weight of a "typical" freshly mixed concrete for different maximum aggregate sizes. The weight of the fine aggregates is determined by subtracting the weight of the other ingredients from the total weight. Since Table 6.10 is based on a "typical" mix, the weight-based mix design method is only approximate.

Table 6.9 Minimum Requirements of Cementing Materials for Concrete Used in Flatwork*

Maximum Size of Aggregate mm(in.)	Cementing Materials, kg/m³ (lb/yd³)**
37.5 (1½)	280 (470)
25.0 (1)	310 (520)
19.0 (¾)	320 (540)
12.5 (½)	350 (590)
9.5 (3/8)	360 (610)

* American Concrete Institute (ACI 302).

** Cementing materials quantities may need to be greater for severe exposure. For example, for deicer exposures, concrete should contain at least 335 kg/m³ (564 lb/yd³) of cementing materials.

Table 6.10 Estimate of Weight of Freshly Mixed Concrete

Maximum Aggregate Size, mm (in.)	Non Air Entrained Concrete kg/m³ (lb/yd³)	Air Entrained Concrete kg/m³ (lb/yd³)
9.5 (3/8)	2276 (3840)	2187 (3690)
12.5 (½)	2305 (3890)	2228 (3760)
19.0 (¾)	2347 (3960)	2276 (3840)
25.0 (1)	2376 (4010)	2311 (3900)
37.5 (1½)	2412 (4070)	2347 (3960)
50.0 (2)	2441 (4120)	2370 (4000)
75.0 (3)	2465 (4160)	2394 (4040)
150 (6)	2507 (4230)	2441 (4120)

In the absolute volume method of mix design, the component weight and the specific gravity are used to determine the volumes of the water, coarse aggregate, and cement. These volumes, along with the volume of the air, are subtracted from a unit volume of concrete to determine the volume of the fine aggregate required. The volume of the fine aggregate is then converted to a weight using the unit weight. Generally, the bulk SSD specific gravity of aggregates is used for the weight-volume conversions of both fine and coarse aggregates.

10. **Moisture Corrections** Mix designs assume that water used to hydrate the cement is the free water in excess of the moisture content of the aggregates at the SSD condition (absorption), as discussed in Chapter 4. Therefore, the final step in the mix design process is to adjust the weight of water and aggregates to account for the existing moisture content of the aggregates. If the moisture content of the aggregates is more than the SSD moisture content, the weight of mixing water is reduced by an amount equal to the free weight of the moisture on the

aggregates. Similarly, if the moisture content is below the SSD moisture content, the mixing water must be increased.

11. Trial Mixes After computing the required amount of each ingredient, a trial batch is mixed to check the mix design. Three 0.15 m × 0.30 m cylinders are made, cured for 28 days, and tested for compressive strength. In addition, the air content and slump of fresh concrete are measured. If the slump, air content, or compressive strength does not meet the requirements, the mixture is adjusted and other trial mixes are made until the design requirements are satisfied.

Additional trial batches could be made by slightly varying the material quantities in order to determine the most workable and economical mix.

Sample Problem 6.3

You are working on a concrete mix design that requires each cubic yard of concrete to have a 0.43 water-cementitious materials ratio, 2077 lb/yd³ of dry gravel, 244 lb/yd³ of water, and 4% air content. The available gravel has a specific gravity of $G_{gravel} = 2.6$, a moisture content of 2.3% and absorption of 4.5%. The available sand has a specific gravity of $G_{sand} = 2.4$, a moisture content of 2.2% and absorption of 1.7%. Air entrainer is to be included using the manufacturers specification of 0.1 fl. oz/1% air/100 lb cement.

For each cubic yard of concrete needed on the job, calculate the weight of cement, moist gravel, moist sand, and water that should be added to the batch. Summarize and total the mix design when finished.

Solution

  Step 7: $W_{cement} = 244/0.43 = 567$ lb/yd³
  Step 8: air entrainer $= (0.1)(4)(567/100) = 2.27$ fl. oz
  Step 9: $\gamma_w = 62.4$ lb/ft³ (3 ft/yd³) $= 1684.8$ lb/yd³
    $V_{cement} = 567.4/(3.15 \times 1684.8) = 0.107$ yd³
    $V_{water} = 244/1684.8 = 0.145$ yd³
    $V_{gravel} = 2077/(2.6 \times 1684.8) = 0.474$ yd³
    $V_{air} = 4\% = 0.040$ yd³
        Subtotal $= 0.766$ yd³
    $V_{sand} = 1 - 0.766 = 0.234$ yd³
    $m_{sand} = (0.234)(2.4)(1684.8) = 946$ lb/yd³
  Step 10: mix water $= 244 - 2077(0.023 - 0.045) - 946.2(0.022 - 0.017) = 285$ lb/yd³
    moist gravel $= 2077(1.023) = 2125$ lb/yd³
    moist sand $= 946.2(1.022) = 967$ lb/yd³
    cement $= 567$ lb/yd³
    air entrainer $= 2.27$ fl. oz

Sample Problem 6.4

Design a concrete mix for the following conditions and constraints using the absolute volume method:

Design Environment
Bridge pier exposed to freezing and subjected to deicing chemicals
Required design strength=24.1 MPa(3500 psi)
Minimum dimension=0.3 m (12 in.)
Minimum space between rebars=50 mm (2 in.)
Minimum cover over rebars=40 mm(1.5 in.)
Standard deviation of compressive strength of 2.4 MPa(350 psi) is expected
  (more than 30 samples)
Only air entrainer is allowed

Available Materials
Cement
  Select Type V due to exposure
Air Entrainer
  Manufacturer specification 6.3 ml/1%air/100 kg cement (0.1 fl oz/1%air/100 lb cement)
Coarse aggregate
  25 mm(1 in.)maximum size, river gravel(round)
  Bulk oven dry specific gravity=2.621, Absorption=0.4%
  Oven dry-rodded density=1681 kg/m³ (105 pcf)
  Moisture content=1.5%
Fine aggregate
  Natural sand
  Bulk oven-dry specific gravity=2.572, Absorption=0.8%
  Moisture content=4%
  Fineness modulus=2.60

Solution

1. Strength Requirements

    s=2.4 MPa(350 psi)(enough samples so that no correction is needed)
    $f'_{cr} = f'_c + 1.34s = 24.1 + 1.34(2.4) = 27.3$ MPa(3960 psi)
    $f'_{cr} = f'_c + 2.33s - 3.45 = 24.1 + 2.33(2.4) - 3.45 = 26.2$ MPa(3810 psi)

    $f'_{cr} = 27.3$ MPa(3960 psi)

2. Water-Cementitious Materials Ratio

    Strength requirement(Table 6.1), water-cementitious materials ratio=0.48 by interpolation
    Exposure requirement(Tables 6.3 and 6.4), maximum water-cementitious materials ratio=0.45

Water-cementitious materials ratio = 0.45

3. Coarse Aggregate Requirements

$$25 \text{ mm} < \frac{1}{5}(300 \text{ mm}) \text{ minimum dimensions}$$

$$25 \text{ mm} < \frac{3}{4}(50 \text{ mm}) \text{ rebar spacing}$$

$$25 \text{ mm} < \frac{3}{4}(40 \text{ mm}) \text{ rebar cover}$$

Aggregate size Okay for dimensions

(Table 6.5) 25 mm maximum size coarse aggregate and 2.60 FM fine aggregate Coarse aggregate factor = 0.69

Dry weight of coarse aggregate = (1681)(0.69) = 1160 kg/m³ (1956 lb/yd³)

Coarse aggregate = 1160 kg/m³ (1956 lb/yd³)

4. Air Content

(Table 6.6) Severe exposure, target air content = 6.0%

Job range = 5% to 8% base

Design using 7%

5. Workability

(Table 6.7) Pier best fits the column requirement in the table

Slump range = 25 to 100 mm (1 to 4 in.)

Use 75 mm (3 in.)

6. Water Content

(Table 6.8) 25 mm aggregate with air entrainment and 75 mm slump

Water = 175 kg/m³ (295 lb/yd³) for angular aggregates. Since we have round coarse aggregates, reduce by 27 kg/m³ (45 lb/yd³)

Required water = 148 kg/m³ (250 lb/yd³)

7. Cementing Materials Content

Water-cementitious materials ratio = 0.45, water = 148 kg/m³ (250 lb/yd³)

Cement = 148/0.45 = 329 kg/m³ (556 lb/yd³)

Increase for minimum criterion of 334 kg/m³ (564 lb/yd³) for exposure

Cement = 334 kg/m³ (564 lb/yd³)

8. Admixture

7% air, cement = 334 kg/m³ (564 lb/yd³)

Admixture = (6.3)(7)(334/100) = 147 ml/m³ (3.9 fl oz/yd³)

Admixture = 147 ml/m³ (3.9 fl oz/yd³)

9. Fine Aggregate Requirements

Find fine aggregate content; use the absolute volume method.

Water volume = 148/(1×1000) = 0.148 m³/m³ (4.006 ft³/yd³)

Cement volume = 334/(3.15×1000) = 0.106 m³/m³ (2.869 ft³/yd³)

Air volume = 0.07 m³/m³ (0.07×27 = 1.890 ft³/yd³)

Coarse aggregate volume $= 1160/(2.621 \times 1000) = 0.443$ m³/m³ (11.960 ft³/yd³)

Subtotal volume $= 0.767$ m³/m³ (20.725 ft³/yd³)

Fine aggregate

volume $= 1 - 0.767 = 0.233$ m³/m³ $(27 - 20.725 = 6.275$ ft³/yd³)

Fine aggregate dry weight $= (0.233)(2.572)(1000) = 599$ kg/m³ (1007 lb/yd³)

Fine aggregate $= 599$ kg/m³ (1007 lb/yd³)

10. Moisture Corrections

Coarse aggregate: Need 1160 kg/m³ (1956 lb/yd³) in dry condition, so increase by 1.5% for moisture

Moist coarse aggregate $= (1160)(1.015) = 1177$ kg/m³ (1985 lb/yd³)

Fine aggregate: Need 599 kg/m³ (1007 lb/yd³) in dry condition, so increase 4%
for moisture

Fine aggregate in moist condition $= (599)(1.04) = 623$ kg/m³ (1047 lb/yd³)

Water: Reduce for free water on aggregates $= 148 - 1160(0.015 - 0.004) - 599(0.04 - 0.008) = 116$ kg/m³ (196 lb/yd³)

	Summary	
	Batch Ingredients Required	
	1 m³ PCC	1 yd³ PCC
Water	116 kg	196 lb
Cement	334 kg	564 lb
Fine aggregate	623 kg	1047 lb
Coarse aggregate	1177 kg	1985 lb
Admixture	147 ml	3.9 fl oz

6.1.2 Mixing Concrete for Small Jobs

The mix design process applies to large jobs. For small jobs, for which a large design effort is not economical (e.g., jobs requiring less than one cubic meter of concrete), Tables 6.11 and 6.12 can be used as a guide. The values in these tables may need to be adjusted to obtain a workable mix, using the locally available aggregates. Recommendations related to exposure conditions discussed earlier should be followed.

Tables 6.11 and 6.12 are used for proportioning concrete mixes by weight and volume, respectively. The tables provide ratios of components, with a sum of one unit. Therefore, the required total weight or volume of the concrete mix can be multiplied by the given ratios to obtain the weight or volume of each component. Note that for proportioning by volume, the combined volume is approximately two-thirds of the sum of the original bulk volumes of the components, since water and fine materials fill the voids between coarse materials.

Table 6.11 Relative Components of Concrete for Small Jobs, by Weight*

Maximum Size of Coarse Aggregate mm (in.)	Air-Entrained Concrete				Non-Air-Entrained Concrete			
	Cement	Wet Fine Aggregate	Wet Coarse Aggregate**	Water	Cement	Wet Fine Aggregate	Wet Coarse Aggregate**	Water
9.5(3/8)	0.210	0.384	0.333	0.073	0.200	0.407	0.317	0.076
12.5(½)	0.195	0.333	0.399	0.073	0.185	0.363	0.377	0.075
19(¾)	0.176	0.296	0.458	0.070	0.170	0.320	0.442	0.068
25(1)	0.169	0.275	0.493	0.063	0.161	0.302	0.470	0.067
37.5(1½)	0.159	0.262	0.517	0.062	0.153	0.287	0.500	0.060

* Portland Cement Association, 2002.
** If crushed stone is used, decrease coarse aggregate by 50 kg and increase fine aggregate by 50 kg for each cubic meter of concrete (or decrease coarse aggregate by 3 lb and increase fine aggregate by 3 lb for each cubic foot of concrete).

Table 6.12 Relative Components of Concrete for Small Jobs, by Volume*

Maximum Size of Coarse Aggregate mm (in.)	Air-Entrained Concrete				Non-Air-Entrained Concrete			
	Cement	Wet Fine Aggregate	Wet Coarse Aggregate	Water	Cement	Wet Fine Aggregate	Wet Coarse Aggregate	Water
9.5(3/8)	0.190	0.429	0.286	0.095	0.182	0.455	0.272	0.091
12.5(½)	0.174	0.391	0.348	0.087	0.167	0.417	0.333	0.083
19(¾)	0.160	0.360	0.400	0.080	0.153	0.385	0.385	0.077
25(1)	0.154	0.346	0.423	0.077	0.148	0.370	0.408	0.074
37.5(1½)	0.148	0.333	0.445	0.074	0.143	0.357	0.429	0.071

* Portland Cement Association, 2002.
The combined volume is approximately 2/3 of the sum of the original bulk volumes.

Sample Problem 6.5

Determine the required weights of ingredients to make a 3500-lb batch of non-air-entrained concrete mix with a maximum gravel size of 1/2 in.

Solution

From Table 6.11:

Weight of cement = 3500 × 0.185 = 647.5 lb

Weight of wet fine aggregate = 3500 × 0.363 = 1270.5 lb

Weight of wet coarse aggregate = 3500 × 0.377 = 1319.5 lb

Weight of water = 3500 × 0.075 = 262.5 lb

Sample Problem 6.6

Determine the required volumes of ingredients to make a 0.5-m³ batch of air-entrained concrete mix with a maximum gravel size of 19 mm.

Solution

Sum of the original bulk volumes of the components = 0.5 × 1.5 = 0.75 m³. From Table 6.12:

Volume of cement = 0.75 × 0.160 = 0.12 m³

Volume of wet fine aggregate = 0.75 × 0.360 = 0.27 m³

Volume of wet coarse aggregate = 0.75 × 0.400 = 0.3 m³

Volume of water = 0.75 × 0.080 = 0.06 m³

6.2 Mixing, Placing, and Handling Fresh Concrete

The proper batching, mixing, and handling of fresh concrete are important prerequisites for strong and durable concrete structures. Next we will discuss the basic steps and precautions to be followed in mixing and handling fresh concrete (Mehta and Monteiro 1993; American Concrete Institute 1982; American Concrete Institute 1983).

Batching is measuring and introducing the concrete ingredients into the mixer. Batching by weight is more accurate than batching by volume, since weight batching avoids the problem created by bulking of damp sand. Water and liquid admixtures, however, can be measured accurately either by weight or volume. On the other hand, batching by volume is commonly used with continuous mixers and when hand mixing.

Concrete should be mixed thoroughly, either in a mixer or by hand, until it becomes uniform in appearance. Hand mixing is usually limited to small jobs or situations in which mechanical mixers are not available. Mechanical mixers include on-site mixers and central mixers in ready-mix plants. The capacity of these mixers varies from 1.5 m³ to 9 m³. Mixers also vary in type, such as tilting, nontilting, and pan-type mixers. Most of the mixers are batch mixers, although some mixers are continuous.

Mixing time and number of revolutions vary with the size and type of the mixer. Specifications usually require a minimum of 1 minute of mixing for stationary mixers of up to 0.75 m³ of capacity, with an increase of 15 seconds for each additional 0.75 m³ of capacity. Mixers are usually charged with 10% of the water, followed by uniform additions of solids and 80% of the water. Finally, the remainder of the water is added to the mixer.

6.2.1 Ready-Mixed Concrete

Ready-mixed concrete is mixed in a central plant, and delivered to the job site in mixing trucks ready for placing. Three mixing methods can be used for ready mixed concrete:

1. Central-mixed concrete is mixed completely in a stationary mixer and delivered in an agitator truck(2 rpm to 6 rpm).
2. Shrink-mixed concrete is partially mixed in a stationary mixer and completed in a mixer truck(4 rpm to 16 rpm).
3. Truck-mixed concrete is mixed completely in a mixer truck (4 rpm to 16 rpm).

Truck manufacturers usually specify the speed of rotation for their equipment. Also, specifications limit the number of revolutions in a truck mixer in order to avoid segregation. Furthermore, the concrete should be discharged at the job site within 90 minutes from the start of mixing, even if retarders are used(ASTM C94).

6.2.2 Mobile Batcher Mixed Concrete

Concrete can be mixed in a mobile batcher mixer at the job site. Aggregate, cement, water, and admixtures are fed continuously by volume, and the concrete is usually pumped into the forms.

6.2.3 Depositing Concrete

Several methods are available to deposit concrete at the jobsite. Concrete should be deposited continuously as close as possible to its final position. Advance planning and good workmanship are essential to reduce delay, early stiffening and drying out, and segregation.

6.2.4 Pumped Concrete

Pumped concrete is frequently used for large construction projects. Special pumps deliver the concrete directly into the forms. Careful attention must be exercised to ensure well-mixed concrete with proper workability. The slump should be between 40 mm to 100 mm before pumping. During pumping, the slump decreases by about 12 mm to 25 mm, due to partial compaction. Blockage could happen during pumping, due to either the escape of water through the voids in the mix or due to friction if fines content is too high(Neville 1981).

6.2.5 Vibration of Concrete

Quality concrete requires thorough consolidation to reduce the entrapped air in the mix. On small jobs, consolidation can be accomplished manually by ramming and tamping the concrete. For large jobs, vibrators are used to consolidate the concrete. Several types of vibrators are available, depending on the application. Internal vibrators are the most common type used on construction projects. These consist of an eccentric weight housed in a spud. The weight is rotated at high speed to produce vibration. The spud is slowly lowered into and through the entire layer of concrete, penetrating into the underlying layer if it is still plastic. The spud is left in place for 5 seconds to 2 minutes, depending on the type of vibrator and the consistency of the concrete. The operator judges the total vibration time required. Over-vibration causes segregation as the mortar migrates to the surface.

Several specialty types of vibrators are used in the production of precast concrete. These include external vibrators, vibrating tables, surface vibrators, electric hammers, and vibratory rollers(Neville 1981).

6.2.6 Pitfalls and Precautions for Mixing Water

Since the water-cementitious materials ratio plays an important role in concrete quality, the water content must be carefully controlled in the field. Water should not be added to the concrete during transportation. Crews frequently want to increase the amount of water in order to improve workability. If water is added, the hardened concrete will suffer serious loss in quality and strength. The engineer in the field must prevent any attempt to increase the amount of mixing water in the concrete beyond that which is specified in the mix design.

6.2.7 Measuring Air Content in Fresh Concrete

Mixing and handling can significantly alter the air content of fresh concrete. Thus, field tests are used to ensure that the concrete has the proper air content prior to placing. Air content can be measured with the pressure, volumetric, gravimetric, or Chace air indicator methods.

The pressure method(ASTM C231)is widely used, since it takes less time than the volumetric method. The pressure method is based on Boyle's law, which relates pressure to volume. A calibrated cylinder(Figure 6.4) is filled with fresh concrete. The vessel is capped and air pressure is applied. The applied pressure compresses the air in the voids of the concrete. The volume of air voids is determined by measuring the amount of volume reduced by the pressure applied. This method is not valid for concrete made with lightweight aggregates, since air in the aggregate voids is also compressed, confounding the measurement of the air content of the cement paste.

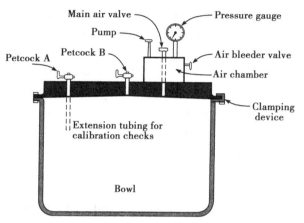

Figure 6.4 Pressure method apparatus for determining air voids in fresh concrete-Type B Meter (ASTM C231). (Copyright ASTM. Reprinted with permission.)

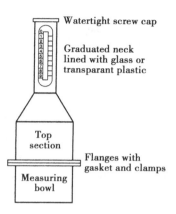

Figure 6.5 Volumetric method (Roll-A-Meter) apparatus for determining air voids in fresh concrete (ASTM C173). (Copyright ASTM. Reprinted with permission.)

The volumetric method for determining air content (ASTM C173) can be used for concrete made with any type of aggregate. The basic process involves placing concrete in a fixed volume cylinder, as shown in Figure 6.5. An equal volume of water is added to the container. Agitation of the container allows the excess water to displace the air in the cement paste voids. The water level in the container falls as the air rises to the top of the container. Thus, the volume of air in the cement paste is directly measured. The accuracy of the method depends on agitating the sample enough to remove all the air from it.

The gravimetric method (ASTM C138) compares the unit weight of freshly mixed concrete with the theoretical maximum unit weight of the mix. The theoretical unit weight is computed from the mix proportions and the specific gravity of each ingredient. This method requires very accurate specific gravity measurements, and thus is more suited to the laboratory rather than the field.

The Chace air indicator test (AASHTO T199) is a quick method used to determine the air content of freshly mixed concrete. The device consists of a small glass tube with a stem, a rub-

Figure 6.6 Chace air indicator.

ber stopper, and a metal cup mounted on the stopper, as shown in Figure 6.6. The metal cup is filled with cement mortar from the concrete to be tested. The indicator is filled with alcohol to a specified level, and the stopper is inserted into the indicator. The indicator is then closed with a finger and gently rolled and tapped until all of the mortar is dispersed in the alcohol and all of the air is displaced with alcohol. With the indicator held in a vertical position, the alcohol level in the stem is read. This reading is then adjusted using calibration tables or figures to determine the air content. The Chace air indicator test can be used to rapidly monitor air content, but it is not precise, nor does it have the repeatability required for specification control. It is especially useful for measuring the air content of small areas near the surface that may have lost air content by improper finishing.

These methods of measuring air content determine the total amount of air, including entrapped air and entrained air, as well as air voids in aggregate particles. Only minute bubbles produced by air-entraining agents impart durability to the concrete. However, the current state of the art is unable to distinguish between the types of air in fresh concrete.

6.2.8 Spreading and Finishing Concrete

Different methods are available to spread and finish concrete, depending on the nature of the structure and the available equipment. Tools and equipment used for spreading and finishing concrete include hand floats, power floats, darbies, bullfloats, straightedges, trowels, vibratory screed, and slip forms.

6.3 Curing Concrete

Curing is the process of maintaining satisfactory moisture content and temperature in the concrete for a definite period of time. Hydration of cement is a long-term process and requires water and proper temperature. Therefore, curing allows continued hydration and, consequently, continued gains in concrete strength. In fact, once curing stops, the concrete dries out, and the strength gain stops, as indicated in Figure 6.7. If the concrete is not cured and is allowed to dry in air, it will gain only about 50% of the strength of continuously cured concrete. If concrete is cured for only three days, it will reach about 60% of the strength of continuously cured concrete; if it is cured for seven days, it will reach 80% of the strength of continuously cured concrete. If curing stops for some time and then resumes again, the strength gain will also stop and reactivate.

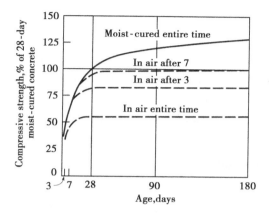

Figure 6.7 Compressive strength of concrete at different ages and curing levels.

Increasing temperature increases the rate of hydration and, consequently, the rate of strength development. Temperatures below 10 °C are unfavorable for hydration and should be avoided, if possible, especially at early ages.

Although concrete of high strength may not be needed for a particular structure, strength is usually emphasized and controlled since it is an indication of the concrete quality. Thus, proper curing not only increases strength, but also provides other desirable properties such as durability, water tightness, abrasion resistance, volume stability, resistance to freeze and thaw, and resistance to deicing chemicals.

Curing should start after the final set of the cement. If concrete is not cured after setting, concrete will shrink, causing cracks. Drying shrinkage can be prevented if ample water is provided for a long period of time. An example of improper curing would be a concrete floor built directly over the subgrade, not cured at the surface, with the moisture in the soil curing it from the bottom. In this case the concrete slab may curl due to the relative difference in shrinkage.

Curing can be performed by any of the following methods:

1. maintaining the presence of water in the concrete during early ages. Methods to maintain the water pressure include pounding or immersion, spraying or fogging, and wet coverings.
2. preventing loss of mixing water from the concrete by sealing the surface. Methods to prevent water loss include impervious papers or plastic sheets, membrane-forming com-

pounds, and leaving the forms in place.
3. accelerating the strength gain by supplying heat and additional moisture to the concrete. Accelerated curing methods include steam curing, insulating blankets or covers, and various heating techniques.

Note that preventing loss of mixing water from the concrete by sealing the surface is not as effective as maintaining the presence of water in the concrete during early ages. The choice of the specific curing method or combination of methods depends on the availability of curing materials, size and shape of the structure, in-place versus plant production, economics, and aesthetics (Kosmatka et al. 2002; American Concrete Institute 1986a).

6.3.1 Ponding or Immersion

Ponding involves covering the exposed surface of the concrete structure with water. Ponding can be achieved by forming earth dikes around the concrete surface to retain water. This method is suitable for flat surfaces such as floors and pavements, especially for small jobs. The method requires intensive labor and supervision. Immersion is used to cure test specimens in the laboratory, as well as other concrete members, as appropriate.

6.3.2 Spraying or Fogging

A system of nozzles or sprayers can be used to provide continuous spraying or fogging. This method requires a large amount of water and could be expensive. It is most suitable in high temperature and low humidity environments. Commercial test laboratories generally have a controlled temperature and humidity booth for curing specimens.

6.3.3 Wet Coverings

Moisture-retaining fabric coverings saturated with water, such as burlap, cotton mats, and rugs are used in many applications. The fabric can be kept wet, either by periodic watering or covering the fabric with polyethylene film to retain moisture. On small jobs, wet coverings of earth, sand, saw dust, hay, or straw can be used. Stains or discoloring of concrete could occur with some types of wet coverings.

6.3.4 Impervious Papers or Plastic Sheets

Evaporation of moisture from concrete can be reduced using impervious papers, such as kraft papers, or plastic sheets, such as polyethylene film. Impervious papers are suitable for horizontal surfaces and simply shaped concrete structures, while plastic sheets are effective and easily applied to various shapes. Periodic watering is not required when impervious papers or plastic sheets are used. Discoloration, however, can occur on the concrete surface.

6.3.5 Membrane-Forming Compounds

Various types of liquid membrane-forming compounds can be applied to the concrete surface to reduce or retard moisture loss. These can be used to cure fresh concrete, as well as to hardened concrete, after removal of forms or after moist curing. Curing compounds can be applied by hand or by using spray equipment. Either one coat or two coats (applied perpendicular to each other) are used. Normally, the concrete surface should be damp when the curing compound is applied. Curing compounds should not be used when subsequent concrete layers are to be placed, since the compound hinders the bond between successive layers. Also, some compounds affect the bond between the concrete surface and paint.

6.3.6 Forms Left in Place

Loss of moisture can be reduced by leaving the forms in place as long as practical, provided that the top concrete exposed surface is kept wet. If wood forms are used, the forms should also be kept wet. After removing the forms, another curing method can be used.

6.3.7 Steam Curing

Steam curing is used when early strength gain in concrete is required or additional heat is needed during cold weather. Steam curing can be attained either with or without pressure. Steam at atmospheric pressure is used for enclosed cast-in-place structures and large precast members. High-pressure steam in autoclaves can be used at small manufactured plants.

6.3.8 Insulating Blankets or Covers

When the temperature falls below freezing, concrete should be insulated using layers of dry, porous material such as hay or straw. Insulating blankets manufactured of fiberglass, cellulose fibers, sponge rubber, mineral wool, vinyl foam, or open-cell polyurethane foam can be used to insulate formwork. Moisture proof commercial blankets can also be used.

6.3.9 Electrical, Hot Oil, and Infrared Curing

Precast concrete sections can be cured using electrical, oil, or infrared curing techniques. Electrical curing includes electrically heated steel forms, and electrically heated blankets. Reinforcing steel can be used as a heatimg element, and concrete can be used as the electrical conductor. Steel forms can also be heated by circulating hot oil around the outside of the structure. Infrared rays have been used for concrete curing on a limited basis.

6.3.10 Curing Period

The curing period should be as long as is practical. The minimum time depends on several factors, such as type of cement, mixture proportions, required strength, ambient weather, size and shape of the structure, future exposure conditions, and method of curing. For most concrete structures the curing period at temperatures above 5 ℃ should be a minimum of seven days or until 70% of specified compressive or flexure strength is attained. The curing period can be reduced to three days if high early strength concrete is used and the temperature is above 10 ℃.

6.4 Properties of Hardened Concrete

It is important for the engineer to understand the basic properties of hardened portland cement concrete and to be able to evaluate these properties. The main properties of hardened concrete that are of interest to civil and construction engineers include the early volume change, creep, permeability, and stress-strain relation.

6.4.1 Early Volume Change

When the cement paste is still plastic it undergoes a slight decrease in volume of about 1%. This shrinkage is known as plastic shrinkage and is due to the loss of water from the cement paste, either from evaporation or from suction by dry concrete below the fresh concrete. Plastic shrinkage may cause cracking; it can be prevented or reduced by controlling water loss.

In addition to the possible decrease in volume when the concrete is still plastic, another form of volume change may occur after setting, especially at early ages. If concrete is not properly cured and is allowed to dry, it will shrink. This shrinkage is referred to as drying shrinkage, and it also causes cracks. Shrinkage takes place over a long period of time, although the rate of shrinkage is high early, then decreases rapidly with time. In fact, about 15% to 30% of the shrinkage occurs in the first two weeks, while 65% to 85% occurs in the first year. Shrinkage and shrinkage-induced cracking are increased by several factors, including lack of curing, high water-cementitious materials ratio, high cement content, low coarse aggregate content, existence of steel reinforcement, and aging. On the other hand, if concrete is cured continuously in water after setting, concrete will swell very slightly due to the absorption of water. Since swelling, if it happens, is very small, it does not cause significant problems. Swelling is accompanied by a slight increase in weight (Neville 1981).

How much drying shrinkage occurs depends on the size and shape of the concrete structure. Also, nonuniform shrinkage could happen due to the nonuniform loss of water. This may happen in mass concrete structures, where more water is lost at the surface than at the interior. In cases such as this, cracks may develop at the surface. In other cases, curling might develop due to the nonuniform curing throughout the structure and, consequently, nonuniform shrinkage.

6.4.2 Creep Properties

Creep is defined as the gradual increase in strain, with time, under sustained load. Creep of concrete is a long-term process, and it takes place over many years. Although the amount of creep in concrete is relatively small, it could affect the performance of structures. The effect of creep varies with the type of structure. In simply supported reinforced concrete beams, creep increases the deflection and, therefore, increases the stress in the steel. In reinforced concrete columns, creep results in a gradual transfer of load from the concrete to the steel. Creep also could result in losing some of the prestress in prestressed concrete structures, although the use of high-tensile stress steel reduces this effect. Rheological models, discussed in Chapter 1, have been used to analyze the creep response of concrete (Neville 1981).

6.4.3 Permeability

Permeability is an important factor that largely affects the durability of hardened concrete. Permeable concrete allows water and chemicals to penetrate, which, in turn, reduces the resistance of the concrete structure to frost, alkali-aggregate reactivity, and other chemical attacks. Water that permeates into reinforced concrete causes corrosion of steel rebars. Furthermore, impervious concrete is a prerequisite in watertight structures, such as tanks and dams.

Typically, the air voids in the cement paste and aggregates are small and do not affect permeability. However, the air voids that do affect permeability of hardened concrete are obtained from two main sources: incomplete consolidation of fresh concrete and voids resulting from evaporation of mixing water that is not used for hydration of cement.

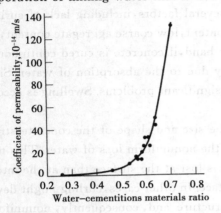

Figure 6.8 Relation between water-cementitious materials ratio and permeability of mature cement paste.

Therefore, increasing the water-cementitious materials ratio in fresh concrete has a severe effect on permeability. Figure 6.8 shows the typical relationship between the water-cementitious materials ratio and the coefficient of permeability of mature cement paste (Powers 1954). It can be seen from the figure that increasing the water-cementitious materials ratio from 0.3 to 0.7 increases the coefficient of permeability by a factor of 1000. For a concrete to be watertight, the water-cementitious materials ratio should not exceed 0.48 for exposure to fresh water and should not be more than 0.44 for exposure to seawater (American Concrete Institute 1975).

Other factors that affect the permeability include age of concrete, fineness of cement particles, and air-entraining agents. Age reduces the permeability, since hydration products fill the spaces between cement grains. The finer the cement particles, the faster is the rate of hydration and the faster is the development of impermeable concrete. Air-entraining agents indirectly reduce the permeability, since they allow the use of a lower water-cementitious materials ratio.

6.4.4 Stress-Strain Relationship

Typical stress-strain behavior of 28-day-old concrete with different water-cementitious materials ratios are shown in Figure 6.9 (Hognestad et al. 1955). It can be seen that increasing the water-cementitious materials ratio decreases both strength and stiffness of the concrete. The figure also shows that the stress-strain behavior is close to linear at low stress levels, then becomes nonlinear as stress increases. With a water-cementitious materials ratio of 0.50 or less and a strain of up to 0.0015, the stress-strain behavior is almost linear. With higher water-cementitious materials ratios, the stress-strain behavior becomes nonlinear at smaller strains. The curves also show that high-strength concrete has sharp peaks and sudden failure characteristics when compared to low-strength concrete.

As discussed in Chapter 1, the elastic limit can be defined as the largest stress that does not cause a measurable permanent strain. When the concrete is loaded slightly beyond the elastic range and then unloaded, a small amount of strain might remain initially, but it may recover

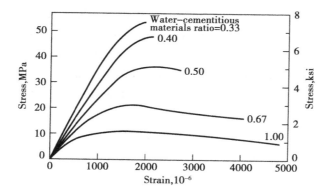

Figure 6.9 Typical stress-strain relations for compressive Tests on 0.15 m×0.30 m concrete cylinders at an age of 28 days.

eventually due to creep. Also, since concrete is not perfectly elastic, the rate of loading affects the stress-strain relation to some extent. Therefore, a specific rate of loading is required for testing concrete. It is interesting to note that the shape of the stress-strain relationship of concrete is almost the same for both compression and tension, although the tensile strength is much smaller than the compressive strength. In fact, the tensile strength of concrete typically is ignored in the design of concrete structures.

The modulus of elasticity of concrete is commonly used in designing concrete structures. Since the stress-strain relationship is not exactly linear, the classic definition of modulus of elasticity (Young's modulus) is not applicable. The initial tangent modulus of concrete has little practical importance. The tangent modulus is valid only for a low stress level where the tangent is determined. Both secant and chord moduli represent "average" modulus values for certain stress ranges. The chord modulus (referred to as the modulus of elasticity) in compression is more commonly used for concrete and is determined according to ASTM C469. The method requires three or four loading and unloading cycles, after which the chord modulus is determined between a point corresponding to a very small strain value and a point corresponding to either 40% of the ultimate stress or a specific strain value. Normal-weight concrete has a modulus of elasticity of 14 GPa to 40 GPa.

Poisson's ratio can also be determined using ASTM C469. Poisson's ratio is used in advanced structural analysis of shell roofs, flat-plate roofs, and mat foundations. Poisson's ratio of concrete varies between 0.11 and 0.21, depending on aggregate type, moisture content, concrete age, and compressive strength. A value of 0.15 to 0.20 is commonly used.

It is interesting to note that both aggregate and cement paste, when tested individually, exhibit linear stress-strain behavior. However, the stress-strain relation of concrete is nonlinear. The reason for this behavior is attributed to the microcracking in concrete at the interface between aggregate particles and the cement paste(Shah and Winter 1968).

The modulus of elasticity of concrete increases when the compressive strength increases, as demonstrated in Figure 6.9. There are several empirical relations between the modulus of elasticity of concrete and the compressive strength. For normal-weight concrete, the relationship used in the United States for designing concrete structures is defined by the ACI Building Code as

$$E_c = 4,731\sqrt{f'_c} \qquad (6.3a)$$

or

$$E_c = 57,000\sqrt{f'_c} \qquad (6.3b)$$

where

E_c = the modulus of elasticity,

f'_c = the compressive strength.

Equation 6.3a is used for SI units, where both E_c and f'_c are in MPa, whereas Equation 6.3b is used for the U.S. customary units, where both E_c and f'_c are in psi. This relation is useful, since it relates the modulus of elasticity(needed for designing concrete structures)with the compressive strength, which can be measured easily in the laboratory.

Sample Problem 6.7

A normal-weight concrete has an average compressive strength of 30 MPa. What is the estimated modulus of elasticity?

Solution

$$E_c = 4731\sqrt{f'_c} = 4731(30)^{1/2} = 25,913 \text{ MPa} = 25.9 \text{ GPa}$$

6.5 Testing of Hardened Concrete

Many tests are used to evaluate the hardened concrete properties, either in the laboratory or in the field. Some of these tests are destructive, while others are nondestructive. Tests can be performed for different purposes; however, they are mostly conducted to control the quality of the concrete and to check specification compliance. Probably the most common test performed on hardened concrete is the compressive strength test, since it is relatively easy to perform and since there is a strong correlation between the compressive strength and many desirable properties (Neville 1981; Mehta and Monteiro 1993). Other tests include split tension, flexure strength, rebound hammer, penetration resistance, ultrasonic pulse velocity, and maturity tests.

6.5.1 Compressive Strength Test

The compressive strength test is the test most commonly performed on hardened concrete. Compressive strength is one of the main structural design requirements to ensure that the structure will be able to carry the intended load. As indicated earlier, compressive strength increases as the water-cementitious materials ratio decreases. Since the water-cementitious materials ratio is directly related to the concrete quality, compressive strength is also used as a measure of quality, such as durability and resistance to weathering. Thus, in many cases, designers specify a high compressive strength of the concrete to ensure high quality, even if this strength is not needed for structural support. The compressive strength f'_c of normal-weight concrete is between 20 MPa to 40 MPa. In the United States, the test is performed on cylindrical specimens and is standardized by ASTM C39. The specimen is prepared, either in the lab or in the field, according to ASTM C192 or C31, respectively. Cores could also be drilled from the structure following ASTM C42. The standard specimen size is 0.15 m in diameter and 0.30 m high, although other sizes with a height-diameter ratio of two can also be used. The diameter of the specimen must be at least three times the nominal maximum size of the coarse aggregate in the concrete.

In the lab, specimens are prepared in three equal layers and are rodded 25 times per layer. After the surface is finished, specimens are kept in the mold for the first 24 ± 8 hours. Specimens are then removed from the mold and cured at 23 ± 1.7 ℃, either in saturated-lime water or in a moist cabinet having a relative humidity of 95% or higher, until the time of testing. Before testing, specimens are capped at the two bases to ensure parallel surfaces. High-strength gypsum plaster, sulfur mortar, or a special capping compound can be used for capping and is applied with a special alignment device (ASTM C617). Using a testing machine, specimens are tested by applying axial compressive load with a specified rate of loading until failure. The compressive strength of the specimen is determined by dividing the maximum load carried by the specimen during the test by the average cross-sectional area. The number of specimens and the number of test batches depend on established practice and the nature of the test program. Usually three or more specimens are tested for each test age and test condition. Test ages often used are 7 days and 28 days.

Note that the test specimen must have a height-diameter ratio of two. The main reason for this requirement is to eliminate the end effect due to the friction between the loading heads and the specimen. Thus, we can guarantee a zone of uniaxial compression within the specimen. If the height-diameter ratio is less than two, a correction factor can be applied to the results as indicated in ASTM C39.

The compressive strength of the specimen is affected by the specimen size. Increasing the specimen size reduces the strength, because there is a greater probability of weak elements where failure starts in large specimens than in small specimens. In general, large specimens have less variability and better representation of the actual strength of the concrete than small specimens. Therefore, the 0.15 m by 0.30 m size is the most suitable specimen size for determining the compressive strength. However, some agencies use 0.10 m diameter by 0.20 m high specimens. The advantages of using smaller specimens are the ease of handling, less possibility of accidental damage, less concrete needed, the ability to use a low-capacity testing machine, and less space needed for curing and storage. Because of the strength variability of small specimens, more specimens should be tested for smaller specimens than are tested for standard-sized specimens. In some cases, five 0.10 m by 0.20 m replicate specimens are used instead of the three replicates commonly used for the standard-sized specimens. Also, when small-sized specimens are used, the engineer should understand the limitations of the test and consider these limitations in interpreting the results.

Figure 6.10 Scanning electron image showing the interface between a sand grain (lower left corner) and the paste.

The interface between the hardened cement paste and aggregate particles is typically the weakest location within the concrete material. When concrete is stressed beyond the elastic range, microcracks develop at the cement paste-aggregate interface and continuously grow until failure(Figure 6.10) shows a scanning electron microscope micrograph of the fractured surface of a hardened cement mortar cylinder at 500x. The figure shows that the cleavage fracture surfaces where sand particles were dislodged during loading. The figure also shows the microcracks around some sand particles developed during loading.

6.5.2 Split-Tension Test

The split-tension test(ASTM C496) measures the tensile strength of concrete. In this test a 0.15 m by 0.30 m concrete cylinder is subjected to a compressive load at a constant rate along the vertical diameter until failure, as shown in Figure 6.11. Failure of the specimen occurs along its vertical diameter, due to tension developed in the transverse direction. The split tensile(indirect tensile) strength is computed as

$$T = \frac{2P}{\pi L d} \quad (6.4)$$

where
  T = tensile strength, MPa(psi),
  P = load at failure, N(psi),
  L = length of specimen, mm(in.), and
  d = diameter of specimen, mm(in.).

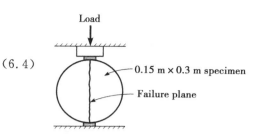

Figure 6.11 Split-tension test.

Typical indirect tensile strength of concrete varies from 2.5 MPa to 3.1 MPa (Neville 1981). The tensile strength of concrete is about 10% of its compressive strength.

6.5.3 Flexure Strength Test

The flexure strength test (ASTM C78) is important for design and construction of road and airport concrete pavements. The specimen is prepared either in the lab or in the field in accordance with ASTM C192 or C31, respectively. Several specimen sizes can be used. However, the sample must have a square cross section and a span of three times the specimen depth. Typical dimensions are 0.15 m by 0.15 m cross section and 0.30 m span. After molding, specimens are kept in the mold for the first 24 ± 8 hours, then removed from the mold and cured at 23 ± 1.7 °C (73.4 ± 3 °F), either in saturated-lime water or in a moist cabinet with a relative humidity of 95% or higher until testing. The specimen is then turned on its side and centered in the third-point loading apparatus, as illustrated in Figure 6.12. The load is continuously applied at a specified rate until rupture. If fracture initiates in the tension surface within the middle third of the span length, the flexure strength (modulus of rupture) is calculated as

$$R = \frac{Mc}{I} = \frac{PL}{bd^2} \quad (6.5)$$

where
  R = flexure strength, MPa(psi),
  M = maximum bending moment = $PL/6$, N.mm(lb.in.),
  $c = d/2$, mm(in.),
  I = moment of inertia = $bh^3/12$, mm^4(in.^{4}),
  P = maximum applied load, which is distributed evenly(1/2 to each) over the two loading points, N(lb),
  L = span length, mm(in.),

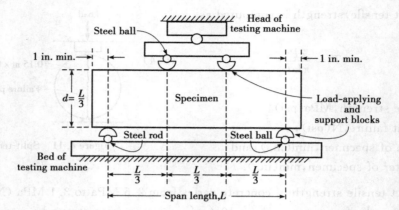

Figure 6.12 Apparatus for flexure test of concrete by third-point loading method (ASTM C78). Copyright ASTM. Reprinted with permission.

b = average width of specimen, mm (in.), and
d = average depth of specimen, mm (in.).

Note that third-point loading ensures a constant bending moment without any shear force applied in the middle third of the specimen. Thus, Equation 6.5 is valid as long as fracture occurs in the middle third of the specimen. If fracture occurs slightly outside the middle third, the results can still be used with some corrections. Otherwise the results are discarded.

For normal-weight concrete, the flexure strength can be approximated as

$$R = (0.62 \text{ to } 0.83)\sqrt{f'_c} \tag{6.6a}$$

$$R = (7.5 \text{ to } 10)\sqrt{f'_c} \tag{6.6b}$$

Equation 6.6a is used for SI units, where both R and f'_c are in MPa, whereas Equation 6.6b is used for U.S. customary units, where both R and f'_c are in psi.

6.5.4 Rebound Hammer Test

The rebound hammer test, also known as the Schmidt hammer test, is a nondestructive test performed on hardened concrete to determine the hardness of the surface (Figure 6.13). The hardness of the surface can be correlated, to some extent, with the concrete strength. The rebound hammer is commonly used to get an indication of the concrete strength. The device is about 0.3 m long and encloses a mass and a spring. The spring-loaded mass is released to hit the surface of the concrete. The mass rebounds, and the amount of rebound is read on a scale attached to the device. The larger the rebound, the harder is the concrete surface and, therefore, the greater is the strength. The device usually comes with graphs prepared by the manufacturer to relate rebound to strength. The test can also be used to check uniformity of the concrete surface.

The test is very simple to run and is standardized by ASTM C805. To perform the test, the hammer must be perpendicular to a clean, smooth concrete surface. In some cases, it would be hard to satisfy this condition. Therefore, correlations, usually provided by the manufacturer, can be used to relate the strength to the amount of rebound at different angles. Rebound hammer results are also affected by several other factors, such as local vibrations, the existence of coarse aggregate particles at the surface, and the existence of voids near the surface. To reduce the effect of these factors, it is desirable to average 10 to 12 readings from different points in the test area.

Figure 6.13 Rebound hammer for nondestructive evaluation of hardened concrete.

6.5.5 Penetration Resistance Test

The penetration resistance test, also known as the Windsor Probe test, is standardized by ASTM C803. The instrument (Figure 6.14) is a gunlike device that shoots probes into the concrete surface in order to determine its strength. The amount of penetration of the probe in the concrete is inversely related to the strength of concrete. The test is almost nondestructive since it creates small holes in the concrete surface.

Figure 6.14 Windsor probe test device.

The device is equipped with a special template with three holes, which is placed on the concrete surface. The test is performed in each of the holes. The average of the penetrations of the three probes through these holes is determined, using a scale and a special plate. Care should be exercised in handling the device to avoid injury. As a way of improving safety, the device cannot be operated without pushing hard on the concrete surface to prevent accidental shooting. The penetration resistance test is expected to provide better strength estimation than the rebound hammer, since the penetration resistance measurement is made not just at the surface but also in the depth of the sample.

6.5.6 Ultrasonic Pulse Velocity Test

The ultrasonic pulse velocity test (ASTM C597) measures the velocity of an ultrasonic wave passing through the concrete (Figure 6.15). In this test, the path length between transducers is divided by the travel time to determine the average velocity of wave propagation. Attempts have been made to correlate pulse velocity data with concrete strength parameters. No good correlations were found, since the relationship between pulse velocity and strength data is affected by a number of variables, such as age of concrete, aggregate-cement ratio, aggregate type, moisture condition, and location of reinforcement (Mehta and Moneiro 1993). This test is used to detect cracks, discontinuities, or internal deterioration in the structure of concrete.

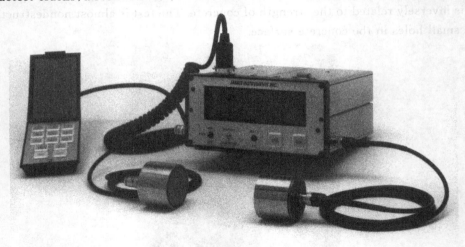

Figure 6.15 Ultrasonic pulse velocity apparatus (Courtesy of James Instrument Inc.).

6.5.7 Maturity Test

Maturity of a concrete mixture is defined as the degree of cement hydration, which varies as a function of both time and temperature. Therefore, it is assumed that, for a particular concrete mixture, strength is a function of maturity. Maturity meters (Figure 6.16) have been developed to provide an estimate of concrete strength by monitoring the temperature of concrete with time. This test (ASTM C1074) is performed on fresh concrete and continued for several days. The maturity meter must be calibrated for each concrete mix.

6.6 Alternatives to Conventional Concrete

There are several alternatives that increase the flexibility and applications of concrete. While a technical presentation of materials for each of these technologies is beyond the scope of this book, the engineer should be aware of some of the materials used to provide additional capabilities of concrete. Some of these alternatives include the following:

- self-consolidating concrete
- flowable fill
- Shotcrete
- lightweight concrete
- high-strength concrete
- shrinkage-compensating concrete
- fiber-reinforced concrete
- heavyweight concrete
- polymers and concrete
- high-performance concrete roller-compacted concrete

6.6.1 Self-Consolidating Concrete

Self-consolidating concrete (SCC), also known as self-compacting concrete, is a highly flowable, nonsegregating concrete that can spread into place, fill the formwork, and encapsulate the reinforcement, without any mechanical consolidation (NRMCA). Some of the advantages of using SCC are the following:

1. It can be placed at a faster rate, with no mechanical vibration and less screeding, resulting

is savings in placement costs.
2. It improves and makes more uniform the architectural surface finish with little or no remedial surface work.
3. It improves ease of filling restricted sections and hard-to-reach areas. This allows the designer to create structural and architectural shapes not achievable with conventional concrete.
4. It improves consolidation around reinforcement and bond with reinforcement.
5. It improves pumpability.

Two important properties specific to self-consolidating concrete in its plastic state are its flowability and stability. The high flowability of SCC is achieved by using high-range water-reducing admixtures without adding extra mixing water. The stability, or resistance to segregation, is attained by increasing the amount of fines and/or by using admixtures that modify the viscosity of the mixture. Fines could be either cementitious materials or mineral fines.

The most common field test that has been used to measure the flowability and stability of SCC is a modified version of the slump test, discussed earlier. In this slump flow test, the slump cone is completely filled, without consolidation. The cone is lifted, and the spread of the concrete is measured, as shown in Figure 6.16. The flowability is measured by the spread, or slump flow. The spread typically ranges from 455 to 810 mm, depending on the requirements of the project. The resistance to segregation, or stability, is measured with the visual stability index (VSI). The VSI is established on the basis of whether bleed water is observed at the leading edge of the spreading concrete or aggregates pile at the center. VSI values range from 0 for "highly stable" to 3 for unacceptable stability.

Figure 6.16 Measuring the spread during the slump flow test.

During the slump flow test, the viscosity can be measured by the rate at which the concrete spreads. The time taken for the concrete to reach a spread diameter of 50 cm from the moment the slump cone is lifted up is measured. This is called the T_{50} (T_{20}) measurement and typically varies between 2 and 10 seconds for SCC. A T_{50} (T_{20}) value indicates a more viscous mix, which is more appropriate for concrete in applications with congested reinforcement or in deep sections. A lower T_{50} (T_{20}) value may be appropriate for concrete that has to travel long horizontal distances without much obstruction.

6.6.2 Flowable Fill

Flowable fill is a self-leveling and self-compacting, cementitious material with an unconfined compressive strength of 8.3 MPa or less. Flowable fill is primarily used as a backfill material in lieu of compacted granular fill. Flowable fill is also commonly referred to as controlled low-strength material(CLSM), controlled density fill (CDF), flowable compacting fill, lean fill, un-shrinkable fill, flow mortar, fly ash flow, and liquid dirt (NRMCA).

The flowable fill mix consists of cement, sand, and water typically mixed with fly ash, ground granulated blast furnace(GGBF) slag, and/or air generating admixtures. The type and proportions of ingredients used to make the flowable fill can largely change its properties to match its intended use.

One of the unique properties of flowable fill that makes it advantageous compared with compacted granular fill is its flowability. High flowability and self-leveling characteristics allow flowable fill to eliminate voids and access spaces that prove to be difficult or impossible with compacted granular fill.

When the flowable fill is placed in the cavity, its initial in-place volume is slightly reduced in the order of 10 mm per meter of depth. This settlement is caused by the displacement of water and the release of entrapped air as a result of consolidation. Once the flowable fill is hardened, the settlement stops and the volume does not change.

Flowable fill can be proportioned to have a wide range of compressive strength. However, the most commonly used flowable fill mixtures are proportioned with consideration of possible excavation in future years and range in compressive strengths between 0.35 MPa and 1 MPa. Flowable fill with a compressive strength less than 150 psi may be excavated manually, while at the same time having adequate bearing capacity to support external loads, such as the weight of a vehicle. Flowable fill exhibiting strengths between 150 and 300 psi will require mechanical equipment for excavation. If the strength exceeds 2 MPa(300 psi), the material is not considered excavatable.

Flowable fill has several advantages over compacted granular backfill. Flowable fill does not require compaction, which is a main concern with granular backfill. Flowable fill can also reach inaccessible locations, such as places around pipes, which are hard to reach with granular backfill. Flowable fill also has a greater bearing capacity than compacted granular fill and no noticeable settlement. It can even be placed in standing water. These advantages result in reduced in-place costs for labor and equipment, as well as time saving during construction.

Flowable fill is typically used as backfill for utility trenches, retaining walls, pipe bedding, and building excavation. It is also used as structural fill for poor soil conditions, mud jacking, floor and footing support, and bridge conversion. Other uses include pavement base, erosion control, and void filling. Flowable fill has been getting more common in recent years and is mostly available at local ready-mix producers.

6.6.3 Shotcrete

Shotcrete is mortar or small-aggregate concrete that is sprayed at high velocity onto a surface. Shotcrete, also known as "gunite" and "sprayed concrete," is a relatively dry mixture that is consolidated by the impact force and can be placed on vertical or horizontal surfaces without sagging.

Shotcrete is applied by either the dry or wet process. In the dry process, a premixed blend of cement and damp aggregate is propelled through a hose by compressed air to a nozzle, while the water is added at the nozzle. In the wet process, all ingredients are premixed and pumped through a hose to the nozzle and forced to the surface using compressed air. In either case, the nozzle should be held perpendicular to the surface to reduce the rebound of coarse aggregates off the surface. The nozzle is held about 0.5 to 1.5 m away from the surface.

Supplementary cementitious materials, such as fly ash and silica fume, can be used in shotcrete to improve workability, chemical resistance, and durability. Accelerating admixtures can also be used to reduce the time of initial set and to allow build up of thicker layers of shotcrete in a single pass. Steel fibers may also be used to improve flexural strength, ductility, and toughness (Kosmatka et al. 2002).

6.6.4 Lightweight Concrete

Students competing in the annual ASCE concrete canoe competition frequently produce concrete with a unit weight less than water. The ACI Guide for the Structural Lightweight Aggregate Concrete requires a 28-day compressive strength of 17 MPa and an air-dried unit weight of less than 1850 kg/m^3 for structural lightweight concrete. The use of lightweight concrete in a structure is usually predicated on the overall cost of the structure; the concrete may cost more, but the reduced dead weight can reduce structural and foundation costs.

The mix proportions for lightweight concrete must compensate for the absorptive nature of the aggregates. Generally, lightweight aggregates are highly absorptive and can continue to absorb water for an extended period of time. This makes the determination of a water-cementitious materials ratio problematical. In addition, the lightweight aggregates tend to segregate by floating to the surface. A minimum slump mix, with air entraining, is used to mitigate this effect.

Nonstructural applications of very lightweight concrete have also been developed. Concrete made with styrofoam "aggregates" has been used for insulation in some building construction.

6.6.5 Heavyweight Concrete

Biological shielding used for nuclear power plants, medical units, and atomic research and test facilities requires massive walls to contain radiation. Concrete is an excellent shielding material. For biological shields, the mass of the concrete can be increased with the use of heavyweight aggregates. The aggregates can be either natural or manufactured. Natural heavyweight aggregates include barite, magnetite, hematite, geothite, illmenite, and ferrophos-phorus. The specific gravity of these aggregates range from 3.4 to 6.5. Steel, with a specific gravity of 7.8, can be used as aggregate for heavyweight concrete. However, the specific gravity of the aggregates makes workability and segregation of heavyweight concrete problematical. Using a higher proportion of sand improves the workability. The workability problem can be avoided by preplacing aggregate, then filling the voids between aggregate particles with grout of cement, sand, water, and admixtures. AsTM C637, *Specifications for Aggregates for Radiation Shielding Concrete* and ASTM C637, *Nomenclature of Constituents of Aggregates for Radiation Shielding Concrete* provide further information on heavyweight concrete practices.

6.6.6 High-Strength Concrete

Concrete made with normal-weight aggregate and having compressive strengths greater than 40 MPa (6000 psi) is considered to be high-strength concrete. Producing a concrete with more than 40 MPa compressive strength requires care in the proportioning of the components and in quality control during construction. The microstructure of concrete with a compressive strength greater than 40 MPa is considerably different from that of conventional concrete. In particular, the porosity of the cement paste and the transition zone between the cement paste and the aggregate are the controlling factors for developing high strength. This porosity is controlled by the water-cementitious materials ratio. The development of superplasticizers has permitted the development of high-strength concrete that is workable and flowable at low water-cementitious materials ratios. In addition, highstrength concrete has excellent durability due to its tight pore structure. In the United States, high-strength concrete is used primarily for skyscrapers. The high-strength and corresponding high elastic modulus allow for reduced structural element size.

6.6.7 Shrinkage-Compensating Concrete

Normal concrete shrinks at early ages, especially if it is not properly cured, as discussed earlier in this chapter. The addition of alumina powders to the cement can cause the concrete to expand at early ages. Shrinkage-compensating cement is marketed as Type K cement. Expansive properties can be used to advantage by restraining the concrete, either by reinforcing or by other means, at early ages. AS the restrained concrete tries to expand, compressive stresses are developed. These compressive stresses reduce the tensile stresses developed by drying shrinkage, and the chance of the concrete cracking due to drying shrinkage is reduced. Details on the design and use of shrinkage-compensating concrete are available in ACI Committee 223 report Recommendations for the Use of Shrinkage-Compensating Cements.

6.6.8 Polymers and Concrete

Polymers can be used in several ways in the production of concrete. The polymer can be used as the sole binding agent to produce polymer concrete. Polymers can be mixed with the plastic concrete to produce polymer-portland cement concrete. Polymers can be applied to hardened concrete to produce polymer-impregnated concrete.

Polymer concrete is a mixture of aggregates and a polymer binder. There are a wide variety of polymers that can be mixed with aggregates to make polymer concrete. Some of these can be used to make rapid-setting concrete that can be put in service in under an hour after placement. Others are formulated for high strength; 140 MPa is possible. Some have good resistance to chemical attack. A common characteristic is that most polymer concretes are expensive, limiting their application to situations in which their unique characteristics make polymer concrete a cost-effective alternative to conventional concrete.

Polymer-portland cement concrete in corporates a polymer into the production of the portland cement concrete. The polymer is generally an elastomeric emulsion, such as latex.

6.6.9 Fiber-Reinforced Concrete

The brittle nature of concrete is due to the rapid propagation of microcracking under applied stress. However, with fiber-reinforced concrete, failure takes place due to fiber pull-out or debonding. Unlike plain concrete, fiber-reinforced concrete can sustain load after initial cracking. This effectively improves the toughness of the material. In addition, the flexural strength of the concrete is improved by up to 30%. For further information on the design and applications of fiber-reinforced concrete, consult the ACI Guide for Specifying, Mixing, Placing and Finishing Steel Fiber Reinforced Concrete.

Fibers are available in a variety of sizes, shapes, and materials (Figure 6.17). The fibers can be made of steel, plastic, glass, and natural materials. Steel fibers are the most common. The shape of fibers is generally described by the aspect ratio, length/diameter. Steel fibers generally have diameters from 0.25 mm to 0.9 mm with aspect ratios of 30 to 150. Glass fiber elements' diameters range from 0.013 mm to 1.3 mm.

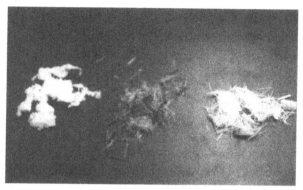

Figure 6.17 Fibers used for fiber-reinforced concrete

The addition of fibers to concrete reduces the workability. The extent of reduction depends on the aspect ratio of the fibers and the volume concentration. Generally, due to construction problems, fibers are limited to a maximum of 2% by volume of the mix. Admixtures can be used to restore some of the workability to the mix.

Since the addition of fibers does not greatly increase the strength of concrete, its use in structural members is limited. In beams, columns, suspended floors, etc., conventional reinforcing must be used to carry the total tensile load. Fiber-reinforced concrete has been successfully used for floor slabs, pavements, slope stabilization, and tunnel linings.

6.6.10 Roller-Compacted Concrete

Based on the unique requirements for mass concrete used for dam construction, roller-compacted concrete(RCC) was developed. This material uses a relatively low cement factor, relaxed gradation requirements, and a water content selected for construction considerations rather than strength. RCC is a no-slump concrete that is transported, placed, and compacted with equipment used for earth and rockfill dam construction. The RCC is hauled by dump trucks, spread with bulldozers, and compacted with vibration compactors. Japanese experience using RCC in construction found several advantages:

1. The mix is economical, because of the low cement content.
2. Formwork is minimal, because of the layer construction method.
3. The low cement factor limits the heat of hydration, reducing the need for external cooling of the structure.
4. The placement costs are lower than those for conventional concrete methods, due to the use of high-capacity equipment and rapid place-ment rates.
5. The construction period is shorter than that for conventional concrete. In addition, experience in the United States has demonstrated that RCC in-place material costs are about one-third those of conventional concrete. The two primary applications of RCC have been for the construction of dams and large paved areas, such as military tank parking aprons.

6.6.11 High-Performance Concrete

While the current specifications for concrete have provided a material that performs reasonably well, there is concern that the emphasis on strength in the mix design process has led to concrete that is inadequate in other performance characteristics. This has led to an interest in developing specifications and design methods for what has been termed high-performance concrete (HPC). The American Concrete Institute (ACI) defines HPC as concrete that meets special performance and uniformity requirements, which cannot always be obtained using conventional ingredients, normal mixing procedures, and typical curing practices. These requirements may include the following enhancements:

ease of placement and compaction
long-term mechanical properties
early-age strength
toughness
volume stability
extended life in severe environments

These enhanced characteristics may be accomplished by altering the aggregate gradation, including special admixtures, and improving mixing and placement practices. Currently, a compressive strength in the order of 70 to 175 MPa can be obtained. As the need for HPC is better understood and embraced by the engineering community, there will probably be a transition in concrete specification from the current prescriptive method to the performance-based or performance-related specifications. A Strategic Highway Research Program (SHRP) study (Zia et al. 1991) defined HPC as

maximum water-cementitious materials ratio of 0.35;
minimum durability factor of 80%, as determined by ASTM C 666,
Procedure A; and minimum strength criteria of either:

a. 21 MPa (3000 psi) within 4 hours after placement (Very Early Strength, VES),
b. 34 MPa (5000 psi) within 24 hours (High Early Strength, HES), or
c. 69 MPa (10,000 psi) within 28 days (Very High Strength, VHS).

Thus, high-performance concrete is characterized by special performance, both short-term and long-term, and uniformity in behavior (Nawy 1996). Such requirements cannot always be achieved by using only conventional materials or applying conventional practices. Since concrete is the most widely used construction material worldwide, new concrete construction has to utilize the currently available new technology in order to eliminate costly future rehabilitation.

Revolutionary new construction materials and modifications and improvements in the behavior of traditional materials have been developed in the last three decades. These developments have been considerably facilitated by increased knowledge of the molecular structure of materials, studies of long-term failures, development of more powerful instrumentation and monitoring techniques, and the need for stronger and better performing materials, suitable for larger structures, longer spans, and more ductility.

In spite of the current advances of the concrete technology and the development of high-performance concretes, it is expected that the concrete industry will continue improving through the development of new components and admixtures, microstructural studies, blended cement compositions, better material selection and proportioning techniques, and more efficient placement techniques.

SUMMARY

The design of durable portland cement concrete materials is the direct responsibility of civil engineers. Selection of the proper proportions of portland cement, water, aggregates, and admixtures, along with good construction practices, dictates the quality of concrete used in structural applications. Using the volumetric mix design method presented in this chapter will lead to concrete with the required strength and durability. However, the proper design of portland cement concrete is irrelevant unless proper construction procedures are followed, including the appropriate mixing, transporting, placing, and curing of the concrete. To ensure that these processes produce concrete with the desired properties, a variety of quality control tests are performed by civil engineers, including slump tests, air content tests, and strength-gain-with-time tests.

While the vast majority of concrete projects are constructed with conventional materials, there are a variety of important alternative concrete formulations available for specialty applications. These alternatives are introduced in this chapter; however, the technology associated with these alternatives is relatively complex, and further study is required in order to fully understand the behavior of these materials.

QUESTIONS AND PROBLEMS

6.1 The design engineer specifies a concrete strength of 5500 psi. Determine the required average compressive strength for
   a. a new plant where s is unknown
   b. a plant where $s=500$ psi for 22 test results
   c. a plant with extensive history of producing concrete with $s=400$ psi
   d. a plant with extensive history of producing concrete with $s=600$ psi

6.2 A project specifies a concrete strength of 24.1 MPa. Materials engineers will design the mix for a strength higher than that to account for variabilities.
   a. Calculate the required average compressive strength of the mix design if the mixing plant has a standard deviation of $s=3.8$ MPa.
   b. Using the ACI code equation, estimate what would be the modulus of elasticity of this concrete at the *required* compressive strength.

6.3 A project specifies a concrete strength of at least 3000 psi. Materials engineers will design the mix for a strength higher than that. Calculate the required average compressive strength of the mix design if the standard deviation is $s=350$ psi. Estimate the modulus of elasticity of the concrete at the required average compressive strength (the calculated strength, not the given strength).

6.4 What is your recommendation for the maximum nominal size of coarse aggregate for the following situation?
   A continuously reinforced concrete pavement cross section contains a layer of No. 6 reinforcing bars at 6-inch centers, such that the steel is just above mid-depth of a 10-inch thick slab. Cover over the top of the steel is therefore about 4 inches.

6.5 A concrete mix with a 3-inch slump, w/c ratio of 0.50, and sand with a fineness modulus of 2.4 contains 1700 lb of coarse aggregate. Compute the required weight of coarse aggregate per cubic yard. To adjust the mix so as to increase the compressive strength, the water-to-cement ratio is reduced to 0.45. Will the quantity of coarse aggregate increase, decrease, or stay the same? Explain your answer.

6.6 Design the concrete mix according to the following conditions:
   Design Environment
   Building frame
   Required design strength = 27.6 MPa
   Minimum dimension = 150 mm

Minimum space between rebar = 40 mm

Minimum cover over rebar = 40 mm

Statistical data indicate a standard deviation of compressive strength of 2.1 MPa is expected (more than 30 samples).

Only air entrainer is allowed.

Available Materials

Air entrainer: Manufacture specification 6.3 ml/1% air/100 kg cement.

Coarse aggregate: 19 mm maximum size, river gravel (rounded)

Bulk oven-dry specific gravity = 2.55, Absorption = 0.3%

Oven-dry rodded density = 1761 kg/m³

Moisture content = 2.5%

Fine aggregate: Natural sand

Bulk oven-dry specific gravity = 2.659, Absorption = 0.5%

Moisture content = 2%

Fineness modulus = 2.47

6.7 The design of a concrete mix requires 1173 kg/m³ of gravel in dry condition, 582 kg/m³ of sand in dry condition, and 157 kg/m³ of free water. The gravel available at the job site has a moisture content of 0.8% and absorption of 1.5%, and the available sand has a moisture content of 1.1% and absorption of 1.3%. What are the masses of gravel, sand, and water per cubic meter that should be used at the job site?

6.8 Design a non-air-entrained concrete mix for a small job with a maximum gravel size of 25 mm (1 in.). Show the results as follows:

a. masses of components to produce 2000 kg (4400 lb) of concrete.

b. volumes of components to produce 1 m³ (36 ft³) of concrete.

6.9 What do we mean by curing concrete? What would happen if concrete is not cured?

6.10 Discuss five different methods of concrete curing.

6.11 Discuss the change in volume of concrete at early ages.

6.12 Discuss the creep response of concrete structures. Provide examples of the effect of creep on concrete structures.

6.13 Comparing PCC with mild steel, answer the following questions:

a. Which one is stronger?

b. Which one has a higher modulus or stiffness?

C. Which one is more brittle?

d. What is the range of compressive strength for a typical PCC?

e. What is the compressive strength for a high-strength concrete?

f. What would be a reasonable range for PCC modulus?

References

American Concrete Institute. Specifications for Structural Concrete. ACI Committee 301 Report, ACI 301-99.

Farmington Hills, MI: American Concrete Institute, 1999.

American Concrete Institute. *Guide for Concrete Floor and Slad Construction*. ACI Committee 302 Report, ACI 302.1R-96. Farmington Hills, MI: American Concrete Institute, 1996.

American Concrete Institute. *Hot-Weather Concreting*. ACI Committee 305 Report, ACI 305R-99. Farmington Hills, MI: American Concrete Institute, 1999.

American Concrete Institute. *Cold-Weather Concreting*. ACI Committee 306 Report, ACI 306R-88. Farmington Hills, MI: American Concrete INstitute, 1997.

American Concrete INstitute. *Standard Practice for Selecting Proportions for Normal, Heavyweight and Mass Concrete*. ACI Committee 211 Report, ACI 211.1-91. Farmington Hills, MI: American Concrete Institute, 1991.

American Concrete Institute. *Standard Practice for Curing Concrete*. ACI Committee 308 Report, ACI 308-92. Farmington Hills, MI: American Concrete Institute, 1997.

American Concrete Institute. *Building Code Requirements for Reinforced Concrete*. ACI Committee 318 Report, ACI 318-99. Farmington Hills, MI: American Concrete Institute, 1999.

Hognestad, E., N. W. Hanson, and D. McHenry. *Concrete Stress Distribution in Ultimate Strength Design*. Development Department Bulletin DX006. Skokie, IL: Portland Cement Association, 1955.

Kosmatka, S. H., B. Kerkhoff, and W. C. Panarese. *Design and Control of Concrete Mixtures*. 14th ed. Skokie, IL: Portland Cement Association, 2002.

Mehta, P. K. and P. J. M. Monteiro. *Concrete Structure, Properties, and Materials*. 2nd ed. Upper Saddle River, NJ: Prentice Hall, 1993.

National Ready Mixed Concrete Association (NRMCA). Concrete in Practice, CIP 37, "Self Consolidating Concrete." Silver Springs, MD: NRMCA, www.nrmca.org

National Ready Mixed Concrete Association (NRMCA). "Flowable Fill Materials." Silver Spring, MD: NRMCA, www.nrmca.org

Mawy, E. G. *Fundamentals of High Strength High Performance Concrete*, Concrete Design and Construction Series. Harlow, UK: Longman Group Limited, 1996.

Neville, A. M. *Properties of Concrete*. 3d ed. London: Pitman Books Ltd, 1981.

Portland Cement Association. *Concrete for Small Jobs*. IS174 T, 1988. Skokie, IL: Portland Cement Association, http://www.portcement.org/pdf_files/IS174

Powers, T. C., L. E. Copeland, J. C. Hayes and H. M. Mann. Permeability of portland cement paste. *Journal of American Concrete Institute* 51(11)285-298, Nov 1954.

Shah, S. P. and G. Winter. Inelastic Behavior and Francture of Concrete. *Symposium on Causes, Mechanism, and Control of Cracking in Concrete*. American Concrete Institute Special Publication no. 20. Farmington Hils, MI: American Concrete Institute, 1968.

Zia, P., M. L. Leming, and S. H. Ahmad. *High Performance Concretes: A State-of-the-Art Report*. SJRP-C/FR-91-103, Washington, DC: Strategic Highway Research Program, National Research Council, 1991. http://www.tfhrc.gov/structur/hpc/hpc2/exec.htm

CHAPTER 7
MASONRY

A masonry structure is formed by combining masonry units, such as stone or brick, with mortar. Masonry is one of the oldest construction materials. Examples of ancient masonry structures include the pyramids of Egypt, the Great Wall of China, and Greek and Roman ruins. Bricks of nearly uniform size became commonly used in Europe during the beginning of the 13th century. The first extensive use of bricks in the United States was around 1600. In the last two centuries, bricks have been used in constructing sewers, bridge piers, tunnel linings, and multistory buildings. Masonry units are still being used in construction in the United States and are competing with other materials, such as wood, steel, and concrete (Adams 1979).

7.1 Masonry Units

Masonry units can be classified as

- concrete masonry units
- clay bricks
- structural clay tiles
- glass blocks
- stone

Concrete masonry units can be either solid or hollow, but clay bricks, glass blocks, and stone are typically solid. Structural clay tiles are hollow units that are larger than clay bricks and are used for lightweight masonry, such as partition walls and filler panels. They can be used with their webs in either a horizontal or a vertical direction. Figure 7.1 shows examples of concrete masonry units, clay bricks, and structural clay tiles. Concrete masonry units and clay bricks are commonly used in the United States.

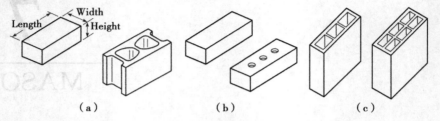

Figure 7.1 Examples of masonry units: (a) concrete masonry units, (b) clay bricks, and (c) structural clay tiles

7.1.1 Concrete Masonry Units

Solid concrete units are commonly called concrete bricks, whereas hollow units are known as concrete blocks, hollow blocks, or cinder blocks. Hollow units have net cross-sectional area in every plane parallel to the bearing surface with less than 75% of the cross-sectional area in the same plane. If this ratio is 75% or more, the unit is categorized as solid (Portland Cement Association 1991).

Concrete masonry units are manufactured in three classes, based on their density: lightweight units, medium-weight units, and normal-weight units, with dry unit weights as shown in Table 7.1. Well-graded sand, gravel, and crushed stone are used to manufacture normal-weight units. Lightweight aggregates such as pumice, scoria, cinders, expanded clay, and expanded shale are used to manufacture lightweight units. Lightweight units are the most common concrete units used in masonry construction because they are easy to handle and transport, and the weight of the structure is reduced. Light-weight units have higher thermal and fire resistance properties and lower sound resistance than normal weight units.

Table 7.1 Weight Classifications and Allowable Maximum Water Absorption of Concrete Masonry Units (ASTM C90 and C129) (Copyright ASTM, reprinted with permission).

Weight Classification	Unit Weight Mg/m³ (pcf)	Maximum Water Absorption kg/m³ (lb/ft³) (Average of 3 units)
Lightweight	1.68(105)	288(18)
Medium Weight	1.68~2.00(105~125)	240(15)
Normal Weight	2.00(125) or more	208(13)

Concrete masonry units are manufactured using a relatively dry (zero-slump) concrete mixture consisting of portland cement, aggregates, water, and admixtures. Type I cement is usually used to manufacture concrete masonry units; however, Type III is sometimes used to reduce the curing time. Air-entrained concrete is sometimes used to increase the resistance of the masonry structure to freeze and thaw effects and to improve workability. compaction, and molding characteristics of the units during manufacturing. The units are molded under pressure, then cured, usually using low-pressure steam curing. Affer manufacturing, the units are stored under controlled conditions so that the concrete continues curing.

Concrete masonry units can be classified as load bearing (ASTM C90) and non-load bearing (ASTM C129). Load-bearing units must satisfy a higher minimum compressive strength requirement than non-load-bearing units, as shown in Table 7.2. The compressive strength of individual concrete masonry units is determined by capping the unit and applying load in the direction of the height of the unit until failure (ASTM C140). A full-size unit is recommended for testing, although a portion of a unit can be used if the capacity of the testing machine is not large enough. The *gross area compressive strength* is calculated by dividing the load at failure by the gross cross-sectional area of the unit. The *net area compressive strength* is calculated by dividing the load at failure by the net cross-sectional area. The net cross-sectional area is calculated by dividing the net volume of the unit by its average height. The net volume is determined using the water displacement method according to ASTM C140.

Table 7.2 Strength Requirements of Concrete Masonry Units (ASTM C90 and C129)

Type	Minimum Compressive Strength Based on Net Area MPa(psi)	
	Average of Three Units	Individual Units
Load bearing	13.1(1900)	11.7(1700)
Non-load-bearing	4.1(600)	3.5(500)

Sample Problem 7.1

A hollow concrete masonry unit has actual gross dimensions of 7-5/8 in. × 7-5/8 in. × 15-5/8 in. The unit is tested in a compression machine with the following results:
  Failure Load = 250 kips
  Net volume of 366.9 in³
  a. Calculate the gross area compressive strength.
  b. Calculate the net area compressive strength.

Solution

  a. Gross area = 7.625 × 15.625 = 119.141 in²

Gross area compressive strength = 250,000/119.141 = 2,098.4 psi
b. Net area = 366.9/7.625 = 48.118 in²
Net area compressive strength = 250,000/48.118 = 5,195.6 psi

The amount of water absorption of concrete masonry units is controlled by ASTM standards to reduce the effect of weathering and to limit the amount of shrinkage due to moisture loss after construction (ASTM C90). The absorption of concrete masonry units is determined by immersing the unit in water for 24 hours (ASTM C140). The absorption and moisture content are calculated as follows.

$$\text{Absorption (kg/m}^3) = \frac{W_s - W_d}{W_s - W_i} \times 1000 \tag{7.1}$$

$$\text{Absorption (lb/ft}^3) = \frac{W_s - W_d}{W_s - W_i} \times 62.4 \tag{7.2}$$

$$\text{Absorption}(\%) = \frac{W_s - W_d}{W_d} \times 100 \tag{7.3}$$

$$\text{Moisture content as a percent of total absorption} = \frac{W_r - W_d}{W_s - W_d} \times 100 \tag{7.4}$$

where
W_s = saturated weight of specimen, kg(lb)
W_d = oven-dry weight of unit, kg(lb),
W_i = immersed weight of specimen, kg(lb), and
W_r = weight of specimen as received

Table 7.1 shows the allowable maximum water absorption for load-bearing concrete masonry units.

Sample Problem 7.2

A concrete masonry unit was tested according to ASTM C140 procedure and produced the following results:

mass of unit as received = 10,354 g
saturated mass of unit = 11,089 g
oven-dry mass of unit = 9893 g

Calculate the percent absorption and moisture content of the unit as a percent of total absorption.

Solution

$$\text{Absorption}(\%) = \frac{11,089 - 9893}{9893} \times 100 = 12.1\%$$

$$\text{Moisture content as a percent of total absorption} = \frac{(10,354 - 9893)}{(11,089 - 9893)} \times 100 = 38.5\%$$

Concrete masonry units are available in different sizes, colors, shapes, and textures. Concrete masonry units are specified by their nominal dimensions. The *nominal dimension* is greater than its specified (or modular) dimension by the thickness of the mortar joint, usually 10 mm. For example, a 200 mm×200 mm×400 mm block has an actual width of 190 mm, height of 190 mm, and length of 390 mm, as illustrated in Figure 7.2. Load-bearing concrete masonry units are available in nominal widths of 100 mm, 150 mm, 200 mm, 250 mm, and 300 mm, heights of 100 mm and 200 mm, and lengths of 300 mm, 400 mm, and 600 mm. Common load-bearing blocks are 200 mm × 200 mm × 400 mm, whereas non-load-bearing blocks are 100 mm×200 mm×400 mm, as shown in Figure 7.3. Also, depending on the position within the masonry wall, they are manufactured as stretcher, single-corner, and double-corner units, as depicted in Figure 7.4.

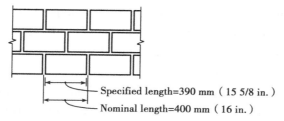

Figure 7.2 Nominal dimensions and specified (modular) dimensions.

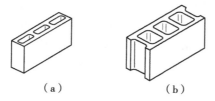

Figure 7.3 Concrete masonry units: (a) Non-load-bearing and (b) load-bearing.

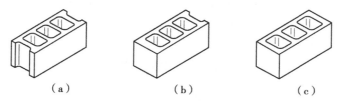

Figure 7.4 Concrete masonry units: (a) stretcher, (b) singl-ecomer, and (c) double-comer.

Solid concrete masonry units (concrete bricks) are manufactured in two grades (N and S) based on strength and absorption requirements. Grade N units have higher compressive strength, resistance to moisture penetration, and resistance to frost action than grade S. According to ASTM C55, the minimum compressive strength of individual units is 20.7 MPa for grade N, and 13.8 MPa for grade S. Grade N bricks are typically used as architectural veneers and facing units in exterior walls. Grade S bricks are for general use where moderate strength and resistance to frost action and moisture penetration are required.

7.1.2 Clay Bricks

Clay bricks are small, rectangular blocks made of fired clay. Clays for brick making vary widely in composition from one place to another. Clays are composed mainly of silica(grains of sand), alumina, lime, iron, manganese, sulfur, and phosphates, with different proportions. Bricks are manufactured by grinding or crushing the clay in mills and mixing it with water to make it plastic. The plastic clay is then molded, textured, dried, and finally fired. Bricks are manufactured with different colors, such as dark red, purple, brown, gray, pink, or dull brown, depending on the firing temperature of the clay during manufacturing. The firing temperature for brick manufacturing varies from 900 ℃ to 1200 ℃ (1650 ℉ to 2200 ℉). Clay bricks have an average density of 2 Mg/m³ (125 pcf).

Bricks are used for different purposes, including building, facing and aesthetics, floor making, and paving. Building bricks(common bricks) are used as a structural material and typically are strong and durable. Facing bricks are used for facing and aesthetic purposes and are available in different sizes, colors, and textures. Floor bricks are used on furnished floor surfaces and are generally smooth and dense and have high resistance to abrasion. Finally, paving bricks are used as a paving material for roads, sidewalks, patios, driveways, and interior floors. Paving bricks are available in different colors, such as red, gray, or brown, and typically they are abrasion resistant and could be vitrified.

Absorption is one of the important properties that determine the durability of bricks. Highly absorptive bricks can cause efflorescence and other problems in the masonry. According to ASTM C67 absorption by 24-hour submersion, absorption by 5-hour boiling, and saturation coefficient are calculated as:

$$\text{Absorption by 24-hour submersion}(\%) = \frac{(W_{s24} - W_d)}{W_d} \times 100 \tag{7.5}$$

$$\text{Absorption by 5-hour boiling}(\%) = \frac{(W_{b5} - W_d)}{W_d} \times 100 \tag{7.6}$$

$$\text{Saturation coefficient} = \frac{(W_{s24} - W_d)}{(W_{b5} - W_d)} \tag{7.7}$$

where

W_d = dry weight of specimen,

W_{s24} = saturated weight after 24-hour submersion in cold water, and

W_{b5} = saturated weight after 5-hour submersion in boiling water.

Clay bricks are very durable, fire resistant, and require very little maintenance. They have moderate insulating properties, which make brick houses cooler in summer and warmer in winter, compared with houses built with other construction materials. Clay bricks are also noncombustible and poor conductors.

The compressive strength of clay bricks is an important mechanical property that controls their load-carrying capacity and durability. The compressive strength of clay bricks is dependent on the composition of the clay, method of brick manufacturing, and the degree of firing. The compressive strength is determined by capping and testing a half unit "flatwise" (load applied in the direction of the height of the unit) and is calculated by dividing the load at failure by the cross-sectional area (ASTM C67). In determining the compressive strength, either the net or gross cross-sectional area is used. Net cross-sectional area is used only if the net cross-section is less than 75% of the gross cross-section. A quarter of a brick can be tested if the capacity of the testing machine is not large enough to test a half brick. Other mechanical properties of bricks include modulus of rupture, tensile strength, and modulus of elasticity. Most clay bricks have modulus of rupture between 3.5 MPa and 26.2 MPa. The tensile strength is typically between 30% to 49% of the modulus of rupture. The modulus of elasticity ranges between 10.3 GPa and 34.5 GPa.

Table 7.3 Physical Requirements for Building Bricks (ASTM C62)
(Copyright ASTM, reprinted with permission).

Grade	Min. Compressive Strength, Gross Area, MPa(psi)		Max. Water Absorption by 5-hour Boiling, %		Max. Saturation Coefficient	
	Average of Five Bricks	Individual	Average of Five Bricks	Individual	Average of Five Bricks	Individual
SW[1]	20.7(3000)	17.2(2500)	17.0	20.0	0.78	0.80
MW[2]	17.2(2500)	15.2(2200)	22.0	25.0	0.88	0.90
NW[3]	10.3(1500)	8.6(1250)	No limit	No limit	No limit	No limit

[1] Severe weathering
[2] Moderate weathering
[3] Negligible weathering

Building bricks are graded according to properties related to durability and resistance to weathering, such as compressive strength, water absorption, and saturation coefficient (ASTM C62). Table 7.3 shows the three available grades and their requirements: SW, MW, and NW, standing for severe weathering, moderate weathering, and negligible weathering, respectively. Grade SW bricks are intended for use in areas subjected to frost action, especially at or below ground level. Grade NW bricks are recommended for use in areas with no frost action and in dry locations, even where subfreezing temperatures are expected. Grade NW bricks can be used in interior construction, where no freezing occurs. Note that higher compressive strengths, lower water absorptions, and lower saturation coefficients are required as the weathering condition gets more severe, so as to reduce the effect of freezing and thawing and wetting and drying.

Sample Problem 7.3

The 5-hour boiling test was performed on a medium weathering clay brick according to ASTM C67 and produced the following masses:

dry mass of specimen = 1.788 kg

saturated mass after 5-hour submersion in boiling water = 2.262 kg

Calculate percent absorption by 5-hour boiling and check whether the brick satisfies the ASTM requirements.

Solution

$$\text{Absorption by 5-hour boiling} = \frac{2.262 - 1.788}{1.788} \times 100 = 26.5\%$$

From Table 7.3, the maximum allowable absorption by 5-hour boiling = 25.0%. Therefore, the brick does not satisfy the ASTM requirements.

Facing bricks (ASTM C216) are manufactured in two durability grades for severe weathering (SW) and moderate weathering (MW). Each durability grade is manufactured in three appearance types: FBS, FBX, and FBA. These three types stand for face brick standard, face brick extra, and face brick architecture. Type FBS bricks are used for general exposed masonry construction. Type FBX bricks are used for general exterior and interior masonry construction, where a high degree of precision and a low permissible variation in size are required. The FBA type bricks are manufactured to produce characteristic architectural effects resulting from nonuniformity in size and texture of the individual units.

Similar to concrete masonry units, bricks are designated by their nominal dimensions. The nominal dimension of the brick is greater than its specified (or modular) dimension by the thickness of the mortar joint, which is about 10 mm and could go up to 12.5 mm. The actual size of the brick depends on the nominal size and the amount of shrinking that occurs during the firing process, which ranges from 4% to 15%.

Clay bricks are specified by their nominal width times nominal height times nominal length. For example, a $4 \times 2\text{-}2/3 \times 8$ brick has nominal width of 100 mm, height of 70 mm, and length of 200 mm. Clay bricks are available in nominal widths ranging from 75 mm to 300 mm, heights from 50 mm to 200 mm, and lengths up to 400 mm. Bricks can be classified as either modular or nonmodular, where modular bricks have widths and lengths of multiples of 100 mm.

7.2 Mortar

Mortar is a mixture of portland cement, lime, sand, and water. Adding a small percentage of lime to the cement mortar makes the mortar "fat" or "rich." which increases its workability. Mortar can be classified as lime mortar or cement mortar. Lime mortar is made of lime, sand, and water, whereas cement (or cement-lime) mortar is made of lime mortar mixed with portland cement (Portland Cement Association 1987).

Mortar is used for the following functions:

- ◆ bonding masonry units together
- ◆ serving as a seating material for the units
- ◆ leveling and seating the units
- ◆ providing aesthetic quality of the structure

Lime mortar gains strength slowly with a typical compressive strength of 0.7 MPa to 2.8 MPa. Cement mortar is mannfactured in four types: M, S, N, and O. Type M has the lowest amount of hydrated lime, whereas type O has the highest amount. The compressive strength of mortar is tested using 50-mm cubes according to ASTM C109. The minimum average compressive strengths of types M, S, N, and O at 28 days are 17.2 MPa, 12.4 MPa, 5.2 MPa, and 2.4 MPa (ASTM C270).

Mortar starts to bind masonry units when it sets. During construction, bricks and blocks should be rubbed and pressed down in order to force the mortar into the pores of the masonry units to produce maximum adhesion. It should be noted, however, that mortar is the weakest part of the masonry wall. Therefore, thin mortar layers generally produce stronger walls than do thick layers.

Unlike concrete, the compressive strength is not the most important property of mortar. Since mortar is used as an adhesive and sealant, it is very important that it forms a complete, strong, and durable bond with the masonry units and with the rebars that might be used to reinforce masonry walls. The ability to bond individual units is measured by the tensile bond strength of mortar (ASTM C952), which is related to the force required to separate the units. The tensile bond strength affects the shear and flexural strength of masonry. The tensile bond strength is usually between 0.14 MPa and 0.55 MPa and is affected by the amount of lime in the mix.

Other properties that affect the performance of mortar are workability, tensile strength, compressive strength, resistance to freeze and thaw, and water retentivity. The water retentivity is a measure, according to ASTM C91, of the rate at which water is lost to the masonry units.

7.3 Grout

Grout is a high-slump concrete consisting of portland cement, sand, fine gravel, water, and sometimes lime. Grout is used to fill the cores or voids in hollow masonry units for the purpose of (1) bonding the masonry units, (2) bonding the reinforcing steel to the masonry, (3) increasing the bearing area, (4) increasing fire resistance, and (5) improving the overturning resistance by increasing the weight. The compressive strength of grout is usually about 14 MPa at 28 days.

7.4 Plaster

Plaster is a fluid mixture of portland cement, lime, sand, and water, which is used for finishing either masonry walls or framed (wood) walls. Plaster is used for either exterior or interior walls. Stucco is plaster used to cover exterior walls. The average compressive strength of plaster is about 13.8 MPa at 28 days.

SUMMARY

Masonry is one of the oldest building technologies, dating back to use of sun-dried adobe blocks in ancient times. Modern masonry units are produced to high standards in the manufacturing process. While the strength of the masonry units is important for quality control, the strength of masonry construction is generally limited by the ability to bond the units together with mortar. The ability of masonry units to resist environmental degradation is an important quality consideration. This ability is closely related to the absorption of the masonry units.

QUESTIONS AND PROBLEMS

7.1 What are the advantages of masonry walls over framed (wood) walls?

7.2 A concrete masonry unit is tested for compressive strength and produces the following results:

Failure load = 593 kN
Gross area = 0.074 m²
Gross volume = 0.014 m³
Net volume = 0.006 m³

Is the unit categorized as solid or hollow? Why? What is the compressive strength? Does the compressive strength satisfy the ASTM requirements for load bearing units shown in Table 7.2?

7.3 A half-block concrete masonry unit is tested for compressive strength. The outside dimensions of the specimen are 7.5 in. × 7.5 in. × 7.5 in. The cross section is a hollow square with a wall thickness of 1 inch. The load is applied perpendicular to the hollow cross section and the maximum load is 46,216 lb.
a. Determine the gross area compressive strength.
b. Determine the net area compressive strength.

7.4 A concrete masonry unit has actual gross dimensions of 7-5/8" × 7-5/8" × 7.5/8". The unit is tested in a compression machine with the following results:

Failure Load = 110 kips
Net volume of 366.2 in³

a. Is the unit categorized as solid or hollow?
b. Calculate the gross area compressive strength.
c. Calculate the net area compressive strength.

7.5 a. Why it is important that, the concrete masonry units meet certain absorption requirements?
b. A portion of a normal-weight concrete masonry unit was tested for absorption and moisture content and produced the following weights:

weight of unit as received = 3.605 lb
saturated weight of unit = 3.939 lb
oven-dry weight of unit = 3.524 lb
immersed weight of unit = 1.684 lb

Calculate the absorption in lb/ft³ and the moisture content of the unit as a percent of total absorption. Does the absorption meet the ASTM C90 requirement for absorption?

7.6 A portion of a medium-weight concrete masonry unit was tested for absorption and moisture content and produced the following results:

mass of unit as received = 5435 g
saturated mass of unit = 5776 g
oven-dry mass of unit = 5091 g
immersed mass of unit = 2973 g

Calculate the absorption in kg/m³ and the moisture content of the unit as a percent of total absorption. Does the absorption meet the ASTM C90 requirement for absorption?

7.7 A portion of a concrete masonry unit was tested for absorption and moisture content according to ASTM C140 procedure and produced the following results:

mass of unit as received = 7805 g
saturated mass of unit = 8223 g
oven-dry mass of unit = 7684 g
immersed mass of unit = 4027 g

Determine the following:
a. percent absorption
b. moisture content of the unit as a percent of total absorption
c. dry density
d. weight classification according to ASTM C90 (lightweight, medium weight, or normal weight).

7.8 What are the functions of mortar?

References

Adams, J. T. *The Complete Concrete, Masonry and Brick Handbook*. New York: Arco, 1979.
Portland Cement Association. *Mortars for Masonry Walls*. Skokie, IL: Portland Cement Association, 1987.
Portland Cement Association. *Masonry Information*. Skokie, IL: Portland Cement Association, 1991.
Somayaji, S. *Civil Engineering Materials*. Upper Saddle River, NJ: Prentice Hall, 2001.

CHAPTER 8

ASPHALT BINDERS AND ASPHALT MIXTURES

Asphalt is one of the oldest materials used in construction. Asphalt binders were used in 3000 B.C., preceding the use of the wheel by 1000 years. Before the mid-1850s, asphalt came from natural pools found in various locations throughout the world, such as the Trinidad Lake asphalt, which is still mined. However, with the discovery and refining of petroleum in Pennsylvania, use of asphalt cement became widespread. By 1907, more asphalt cement came from refineries than came from natural deposits. Today, practically all asphalt cement is from refined petroleum.

Bituminous materials are classified as asphalts and tars, as shown in Figure 8.1. Several asphalt products are used; asphalt is used mostly in pavement construction, but is also used as sealing and waterproofing agents. Tars are produced by the destructive distillation of bituminous coal or by cracking petroleum vapors. In the United States, tar is used primarily for waterproofing membranes, such as roofs. Tar may also be used for pavement treatments, particularly where fuel spills may dissolve asphalt cement, such as on parking lots and airport aprons.

The fractional distillation process of crude petroleum is illustrated in Figure 8.2. Different products are separated at different temperatures. Figure 8.2 shows the main products, such as gasoline, kerosene, diesel oil, and asphalt residue (asphalt cement). Since asphalt is a lower-valued product than other components of crude oil, refineries are set up to produce the more valuable fuels at the expense of asphalt production. The quantity and quality of the asphalt depends on the crude petroleum source and the refining method. Some crude sources, such as the Nigerian oils, produce little asphalt, while others, such as many of the Middle Eastern oils, have a high asphalt content.

204 Materials for Civil and Construction Engineers

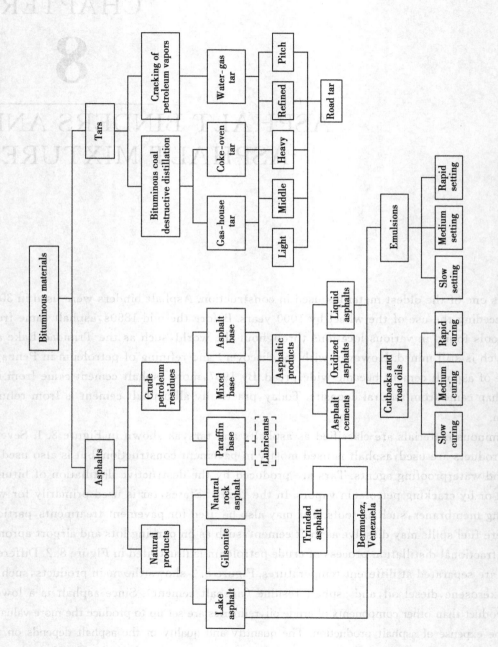

Figure 8.1 Classification of bituminous materials. (Goetz and Wood 1960).

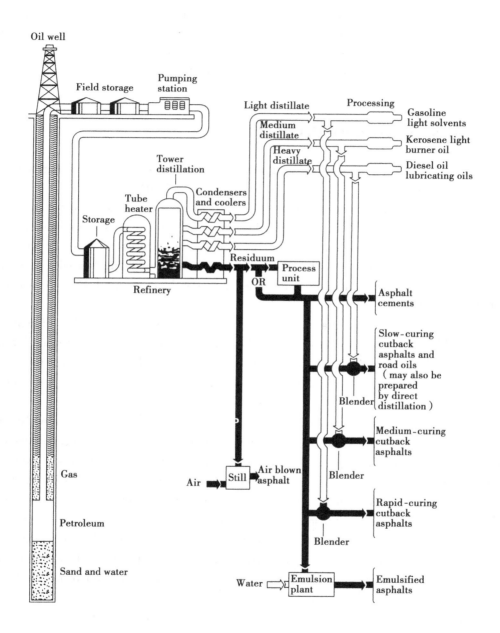

Figure 8.2 Distillation of crude petroleum. (The Asphalt Institute 1989).

This chapter reviews the types, uses, and chemical and physical properties of asphalt. The asphalt concrete used in road and airport pavements, which is a mixture of asphalt and aggregates, is also presented. The chapter discusses the recently developed Performance Grade asphalt binder specifications and Superpave mix design. Recycling of pavement materials and additives used to modify the asphalt properties are also included.

8.1 Types of Asphalt Products

Asphalt used in pavements is produced in three forms: asphalt cement, asphalt cutback, and asphalt emulsion. Asphalt cement is a blend of hydrocarbons of different molecular weights. The characteristics of the asphalt depend on the chemical composition and the distribution of the molecular weight hydrocarbons. As the distribution shifts toward heavier molecular weights, the asphalt becomes harder and more viscous. At room temperatures, asphalt cement is a semisolid material that cannot be applied readily as a binder without being heated. Liquid asphalt products, cutbacks and emulsions, have been developed and can be used without heating (The Asphalt Institute 1989).

Although the liquid asphalts are convenient, they cannot produce a quality of asphalt concrete comparable to what can be produced by heating neat asphalt cement and mixing it with carefully selected aggregates. Asphalt cement has excellent adhesive characteristics, which make it a superior binder for pavement applications. In fact, it is the most common binder material used in pavements.

A cutback is produced by dissolving asphalt cement in a lighter molecular weight hydrocarbon solvent. When the cutback is sprayed on a pavement or mixed with aggregates, the solvent evaporates, leaving the asphalt residue as the binder. In the past, cutbacks were widely used for highway construction. They were effective and could be applied easily in the field. However, three disadvantages have severely limited the use of cutbacks. First, as petroleum costs have escalated, the use of these expensive solvents as a carrying agent for the asphalt cement is no longer cost effective. Second, cutbacks are hazardous materials due to the volatility of the solvents. Finally, application of the cutback releases environmentally unacceptable hydrocarbons into the atmosphere. In fact, many regions with air pollution problems have outlawed the use of any cutback material.

An alternative to dissolving the asphalt in a solvent is dispersing the asphalt in water as emulsion. In this process the asphalt cement is physically broken down into micron-sized globules that are mixed into water containing an emulsifying agent. Emulsified asphalts typically consist of about 60% to 70% asphalt residue, 30% to 40% water, and a fraction of a percent of emulsifying agent. There are many types of emulsifying agents; basically they are a soap material. The emulsifying molecule has two distinct components, the head portion, which has an electrostatic charge, and the tail portion, which has a high affinity for asphalt. The charge can be either positive to produce a cationic emulsion or negative to produce an anionic emulsion. When asphalt is introduced into the water with the emulsifying agent, the tail portion of the emulsifier attaches itself to the asphalt, leaving the head exposed. The electric charge of the emulsifier causes a repulsive force between the asphalt globules, which maintains their separation in the water. Since the specific gravity of asphalt is very near that of water, the globules have a neutral buoyancy and, therefore, do not tend to float or sink. When the emulsion is mixed with aggregates or used on a pavement, the water evaporates, allowing the asphalt globs to come together, forming the binder. The phenomenon of separation between the asphalt residue and water is referred to as breaking or setting. The rate of emulsion setting can be controlled by varying the type and amount of the emulsifying agent.

Since most aggregates bear either positive surface charges (such as limestone) or negative surface charges (such as siliceous aggregates), they tend to be compatible with anionic or cationic emulsions, respectively. However, some emulsion manufacturers can produce emulsions that bond well to aggregate-specific types, regardless of the surface charges.

Although emulsions and cutbacks can be used for the same applications, the use of emulsions is increasing because they do not include hazardous and costly solvents.

8.2 Uses of Asphalt

The main use of asphalt is in pavement construction and maintenance. In addition, asphalt is used in sealing and waterproofing various structural components, such as roofs and underground foundations.

The selection of the type and grade of asphalt depends on the type of construction and the climate of the area. Asphalt cements, also called asphalt binders, are used typically to make hot-mix asphalt concrete for the surface layer of asphalt pavements. Asphalt concrete is also used in patching and repairing both asphalt and portland cement concrete pavements. Liquid asphalts (emulsions and cutbacks) are used for pavement maintenance applications, such as fog seals, chip seals, slurry seals, and microsurfacing (The Asphalt Institute 1989, Mamlouk and Zaniewski 1998). Liquid asphalts may also be used to seal the cracks in pavements. Liquid asphalts are mixed with aggregates to produce cold mixes, as well. Cold mixtures are normally used for patching (when hot-mix asphalt concrete is not available), base and subbase stabilization, and surfacing of low-volume roads. Table 8.1 shows common paving applications for asphalts.

Table 8.1 Paving Applications of Asphalt

Term	Description	Application
Hot mix asphalt	Carefully designed mixture of asphalt and aggregates	Pavement surface, patching
Cold mix	Mixture of aggregates and liquid asphalt	Patching, low volume road surface, asphalt stabilized base
Fog seal	Spray of diluted asphalt emulsion on existing pavement surface	Seal existing pavement surface
Prime coat	Spray coat to bond aggregate base and asphalt concrete surface	Construction of flexible pavement
Tack coat	Spray coat between lifts of asphalt concrete	Construction of new pavements or between an existing pavement and an overlay
Chip seal	Spray coat of asphalt emulsion (or asphalt cement or cutback) followed with aggregate layer	Maintenance of existing pavement or low volume road surfaces
Slurry seal	Mixture of emulsion, well-graded fine aggregate and water	Resurface low volume roads
Microsurfacing	Mixture of polymer modified emulsion, well-graded crushed fine aggregate, mineral filler, water, and additives	Texturing, sealing, crack filling, rut filling, and minor leveling

8.3 Temperature Susceptibility of Asphalt

The consistency of asphalt is greatly affected by temperature. Asphalt gets hard and brittle at low temperatures and soft at high temperatures. Figure 8.3 shows a conceptual relation between temperature and logarithm of viscosity. The viscosity of the asphalt decreases when the temperature increases. Asphalt's temperature susceptibility can be represented by the slope of the line shown in Figure 8.3. The steeper the slope the higher the temperature susceptibility of the asphalt. However, additives can be used to reduce this susceptibility.

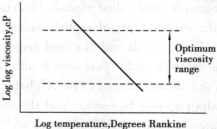

Figure 8.3 Typical relation between asphalt viscosity and temperature.

When asphalt is mixed with aggregates, the mixture will perform properly only if the asphalt viscosity is within an optimum range. If the viscosity of asphalt is higher than the optimum range, the mixture will be too brittle and susceptible to low-temperature cracking(Figure 8.4). On the other hand, if the viscosity is below the optimum range, the mixture will flow readily, resulting in permanent deformation(rutting).

Figure 8.4 Thermal cracking resulting from the use of too stiff asphalt in a cold climate area.

Due to temperature susceptibility, the grade of the asphalt cement should be selected according to the climate of the area. The viscosity of the asphalt should be mostly within the optimum range for the area's annual temperature range; soft-grade asphalts are used for cold climates and hard-grade asphalts for hot climates (See Figure 8.5).

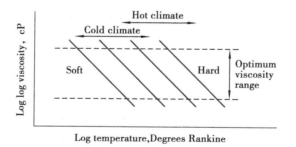

Figure 8.5 Selecting the proper grade of asphalt binder to match the climate.

8.4 Chemical Properties of Asphalt

Asphalt is a mixture of a wide variety of hydrocarbons primarily consisting of hydrogen and carbon atoms, with minor components such as sulfur, nitrogen, and oxygen (heteroatoms), and trace metals. The percentages of the chemical components, as well as the molecular structure of asphalt, vary depending on the crude oil source (Peterson 1984).

The molecular structure of asphalt affects the physical and aging properties of asphalt, as well as how the asphalt molecules interact with each other and with aggregate. Asphalt molecules have three arrangements, depending on the carbon atom links: (1) aliphatic or parraffinic, which form straight or branched chains, (2) saturated rings, which have the highest hydrogen to carbon ratio, and (3) unsaturated rings or aromatic. Heteroatoms attached to carbon alter the molecular configuration. Since the number of molecular structures of asphalt is extremely large, research on asphalt chemistry has focused on separating asphalt into major fractions that are less complex or more homogeneous. Each of these fractions is a complex chemical structure.

Asphalt cement consists of asphaltenes and maltenes (petrolenes). The maltenes consist of resins and oils. The asphaltenes are dark brown friable solids that are chemically complex, with the highest polarity among the components. The asphaltenes are responsible for the viscosity and the adhesive property of the asphalt. If the asphaltene content is less than 10%, the asphalt concrete will be difficult to compact to the proper construction density. Resins are dark and semisolid or solid, with a viscosity that is largely affected by temperature. The resins act as agents to disperse asphaltenes in the oils; the oils are clear or white liquids. When the resins are oxidized, they yield asphaltene-type molecules. Various components of asphalt interact with each other to form a balanced or compatible system. This balance of components makes the asphalt suitable as a binder.

Three fractionation schemes are used to separate asphalt components, as illustrated in Figure 8.6. The first scheme [Figure 8.6(a)] is partitioning with partial solvents in which n-butranol is added to separate (precipitate) the asphaltics. The butranol is then evaporated, and the remaining component is dissolved in acetone and chilled to -23 ℃ to precipitate the paraffinics and leave the cyclics in solution. The second scheme [Figure 8.6(b)] is selective adsorption-desorption, in which n-heptane is added to separate asphaltene. The remaining maltine fraction is introduced to a chromatographic column and desorbed using solvents with increasing polarity to separate other fractions. The third scheme [Figure 8.6(c)] is chemical precipitation in which *n*-pentane is added to separate the asphaltenes. A sulfuric acid (H_2SO_4) is added in increasing strengths to precipitate other fractions.

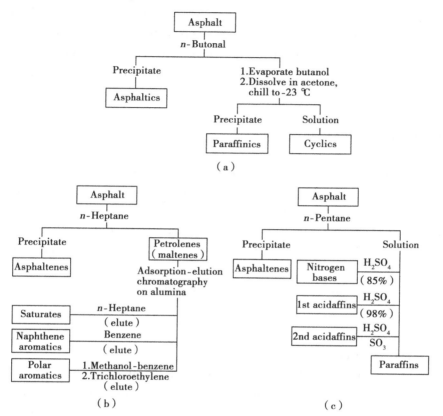

Figure 8.6 Schematic diagrams of three asphalt fractionation schemes: (a) partitioning with partial solvents, (b) selective-adsorbtion-description, and (c) chemical precipitation (Peterson 1984).

In addition, asphalt can be separated based on the molecular size with the use of high-pressure liquid chromatography (gel-permeation chromatography).

8.5 Superpave and Performance Grade Binders

In 1987, the Strategic Highway Research Program (SHRP) began developing a new system for specifying asphalt materials and designing asphalt mixes. The SHRP research program produced the Superpave (Superior Performing Asphalt Pavements) mix design method for asphalt concrete and the Performance Grading method for asphalt binder specification (McGennis

1994;1995). The objectives of SHRP's asphalt research were to extend the life or reduce the life-cycle costs of asphalt pavements, to reduce maintenance costs, and to minimize premature failures. An important result of this research effort was the development of performance-based specifications for asphalt binders and mixtures to control three distress modes: rutting, fatigue cracking, and thermal cracking. Note that the Performance Grade specifications use the term asphalt binder, which refers to asphalt cement with or without the addition of modifiers.

8.6 Asphalt Concrete

Asphalt concrete, also known as hot-mix asphalt (HMA), consists of asphalt cement and aggregates mixed together at a high temperature and placed and compacted on the road while still hot. Asphalt (flexible) pavements cover approximately 93% of the 2 million miles of paved roads in the United States, while the remaining 7% of the roads are portland cement concrete (rigid) pavements. The performance of asphalt pavements is largely a function of the asphalt concrete surface material.

The objective of the asphalt concrete mix design process is to provide the following properties (Roberts et al. 1996):

1. stability or resistance to permanent deformation under the action of traffic loads, especially at high temperatures
2. fatigue resistance to prevent fatigue cracking under repeated loadings
3. resistance to thermal cracking that might occur due to contraction at low temperatures
4. resistance to hardening or aging during production in the mixing plant and in service
5. resistance to moisture-induced damage that might result in stripping of asphalt from aggregate particles
6. skid resistance, by providing enough texture at the pavement surface
7. workability, to reduce the effort needed during mixing, placing and compaction

Regardless of the set of criteria used to state the objectives of the mix design process, the design of asphalt concrete mixes requires compromises. For example, extremely high stability often is obtained at the expense of lower durability, and vice versa. Thus, in evaluating and adjusting a mix design for a particular use, the aggregate gradation and asphalt content must strike a favorable balance between the stability and durability requirements. Moreover, the produced mix must be practical and economical.

8.7 Asphalt Concrete Mix Design

The purpose of asphalt concrete mix design is to determine the design asphalt content using the available asphalt and aggregates. The design asphalt content varies for different material types, material properties, loading levels, and environmental conditions. To produce good-quality asphalt concrete, it is necessary to accurately control the design asphalt content. If the appropriate design asphalt content is not used, the pavement will lack durability or stability, resulting in premature pavement failure. For example, if not enough asphalt binder is used, not all the aggregate particles will be coated with asphalt, which will result in a less stable and less durable material. Also, if too much binder is used, aggregate particles may have too much "lubrication" and may move relative to each other upon application of the load, resulting in a less stable material. Typical design asphalt contents range from 4% to 7% by weight of total mix.

Before the Superpave mix design method was developed during the SHRP program, there were two common asphalt concrete design methods: Marshall (ASTM D1559), and Hveem (ASTM D1560). The Marshall method was more commonly used than the Hveem method, due to its relative simplicity and its ability to be used for field control. Both methods are empirical in nature; that is, they are based on previous observations. Both methods have been used satisfactorily for several decades and have produced long-lasting pavement sections. However, due to their empirical nature they are not readily adaptable to new conditions, such as modified binders, large-sized aggregates, and heavier traffic loads.

The Superpave design system is performance based and is more rational than the Marshall and Hveem methods. Many highway agencies are implementing the Superpave system.

8.7.1 Specimen Preparation in the Laboratory

Asphalt concrete specimens are prepared in the laboratory for mix-design and quality-control tests. To prepare specimens in the lab, aggregates are batched and heated, according to a specified gradation. Asphalt cement is also heated separately and added to the aggregate at a specified rate. Aggregates and asphalt are mixed with a mechanical mixer until the aggregate particles are completely coated with asphalt. Three compaction machines are commonly used:

1. Superpave gyratory compactor
2. Marshall hammer
3. California kneading compactor

Regardless of the compaction method, the procedure for preparing specimens basically follows the same four steps:

1. Heat and mix the aggregate and asphalt cement
2. Place the material into a mold
3. Apply compactive force
4. Allow the specimen to cool and extrude from the mold

The specific techniques for placing the material into the mold vary among the three compaction methods, and the standards for the test must be followed.

The greatest difference among the compaction procedures is the manner in which the compactive force is applied. For the gyratory compaction, the mixture in the mold is placed in the compaction machine at an angle to the applied force. As the force is applied the mold is gyrated, creating a shearing action in the mixture. Gyratory compaction devices have been available for a long time, but their use was limited due to the lack of a mix-design procedure based on this type of compaction. However, the Superpave mix design method (FHWA, 1995) uses a gyratory compactor; thus, this compaction method is now common. Figure 8.7 shows the Superpave gyratory compactor.

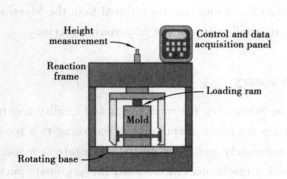

Figure 8.7 Superpave gyratory compactor.

Figure 8.8 Marshall compactor.

In the Marshall procedure (Figure 8.8), a slide hammer weighing 4.45 kg (10 lb) is dropped from a height of 0.46 m to create an impact compaction force (ASTM D1559). The head of the Marshall hammer has a diameter equal to the specimen size, and the hammer is held flush with the specimen at all times.

In the California kneading compactor method (Figure 8.9), the area of the compactor foot is smaller than the area of the mold. After each compaction stroke the mold is rotated, subjecting the asphalt mixture to a kneading action (ASTM D1561). After the kneading compaction is complete, the specimen is reheated while still in the mold; then a compression machine is used to apply a static force to level the face of the specimen.

The Superpave gyratory compactor is used for the Superpave mix design, whereas the Marshall hammer and the California kneading compactor are used for the Marshall and Hveem methods of mix design, respectively. The Superpave gyratory compactor produces specimens 150 mm in diameter and 95 mm to 115 mm high, allowing the use of aggregates with a maximum size of more than 25 mm. Specimens prepared with both Marshall and California kneading compactors, as well as some gyratory compactors, are typically 101.6 mm in diameter and 63.5 mm high.

Figure 8.9 California kneading compactor.

8.7.2 Density and Voids Analysis

It is important to understand the density and voids analysis of compacted asphalt mixtures for both mix design and construction control. Regardless of the method used, the mix design is a process to determine the volume of asphalt binder and aggregates required to produce a mixture with the desired properties. However, since volumes are difficult and not practical to measure, weights are used instead; the specific gravity is used to convert from weight to volume. Figure 8.10 shows that the asphalt mixture consists of aggregates, asphalt binder, and air voids. Note that a portion of the asphalt is absorbed by aggregate particles. Three important parameters commonly used are percent of air voids (voids in total mix) (VTM), voids in the

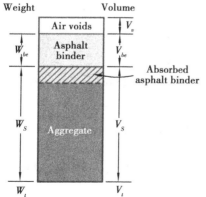

Figure 8.10 Components of compacted asphalt mixture.

mineral aggregate(VMA), and voids filled with asphalt(VFA). These are defined as

$$\text{VTM} = \frac{V_v}{V_m} 100 \tag{8.1}$$

$$\text{VMA} = \frac{V_v + V_{be}}{V_m} 100 \tag{8.2}$$

$$\text{VFA} = \frac{V_{be}}{V_{be} + V_v} 100 \tag{8.3}$$

where

V_v = volume of air voids
V_{be} = volume of effective asphalt binder
V_m = total volume of the mixture

The effective asphalt is the total asphalt minus the absorbed asphalt.

Sample Problem 8.1

A compacted asphalt concrete specimen contains 5% asphalt binder(Sp. Gr. 1.023) by weight of total mix, and aggregate with a specific gravity of 2.755. The bulk density of the specimen is 2.441 Mg/m³. Ignoring absorption, compute VTM, VMA, and VFA.

Solution

Referring to Figure SP 8.1, assume $V_t = 1$ m³
Determine mass of mix and components:

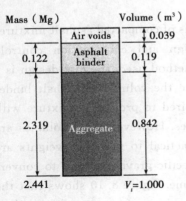

Figure SP8.1

Total mass = 1 × 2.441 = 2.441 Mg
Mass of asphalt binder = 0.05 × 2.441 = 0.122 Mg
Mass of aggregate = 0.95 × 2.441 = 2.319 Mg

Determine volume of components:

$$V_b = \frac{0.122}{1.023} = 0.119 \text{ m}^3$$

Since the problem statement specified no absorption, $V_{be} = V_b$

$$V_s = \frac{2.319}{2.755} = 0.842 \text{ m}^3$$

V_s = Volume of stone (aggregate)

Determine volume of voids:

$$V_v = V_m - V_b - V_s = 1 - 0.199 - 0.842 = 0.039 \text{ m}^3$$

Volumetric calculations:

$$\text{VTM} = \frac{V_v}{V_m} 100 = \frac{0.039}{1.00} 100 = 3.9\%$$

$$\text{VMA} = \frac{V_v + V_{be}}{V_m} 100 = \frac{0.039 + 0.119}{1.00} 100 = 15.8\%$$

$$\text{VFA} = \frac{V_{be}}{V_{be} + V_v} 100 = \frac{0.119}{0.119 + 0.039} 100 = 75\%$$

The density and void analysis requires using the effective specific gravity of the asphalt-coated aggregate, determined from the theoretical maximum specific gravity of the mix (ASTM D2041). The weight of the loose mixture specimen in air A is measured along with the weight of the measurement bowl filled with water D and the weight of the bowl containing the asphalt mix and filled with water E. When the loose mixture specimen is submerged in water, a vacuum is used to remove all air from the sample. The theoretical maximum specific gravity is

$$G_{mm} = \frac{A}{A + D - E} \tag{8.4}$$

It is necessary to determine only the theoretical maximum specific gravity of the sample at one asphalt content. However, the result should be based on the average of three samples (with a minimum of two). By definition, the theoretical maximum specific gravity of asphalt concrete is

$$G_{mm} = \frac{100}{\left(\dfrac{P_s}{G_{se}} + \dfrac{P_b}{G_b}\right)} \tag{8.5}$$

Solving this equation for G_{se} produces

$$G_{se} = \frac{P_s}{\left(\dfrac{100}{G_{mm}} - \dfrac{P_b}{G_b}\right)} \tag{8.6}$$

where

G_{mm} = theoretical maximum specific gravity of the asphalt concrete
P_s = percent weight of the aggregate
P_b = percent weight of the asphalt cement
G_{se} = effective specific gravity of aggregate coated with asphalt
G_b = specific gravity of the asphalt binder

Although G_{se} is determined for only one asphalt content, it is constant for all asphalt contents. Thus, once G_{se} is determined based on the results of the theoretical maximum specific gravity test, it call be used in Equation 8.5 to calculate G_{mm} for the different asphalt contents.

The next step in the process is to determine the bulk specific gravity G_{mb} (ASTM D2726) of each of the compacted specimens. This requires weighing the sample in three conditions: dry, saturated-surface dry, and submerged. The bulk specific gravity is computed as

$$G_{mb} = \frac{\text{Weight in air}}{(\text{Weight SSD} - \text{Weight in water})} \tag{8.7}$$

The unit weight of each specimen is computed by multiplying the bulk specific gravity by the density of water, 1 Mg/m³ (62.4 lb/ft³). The average bulk specific gravity and unit weight for each asphalt content are computed and used to calculate VTM as follows:

$$\text{VTM} = 100\left(1 - \frac{G_{mb}}{G_{mm}}\right) \tag{8.8}$$

The percent voids in the mineral aggregate (VMA), is a measure of the space available in the aggregates for the addition of the asphalt cement. The percent VMA is the volume of the mix minus the volume of the aggregates, divided by the volume of the mix and converted to a percent. VMA is commonly computed from the bulk specific gravity of the aggregate G_{sb}, the bulk specific gravity of the mix G_{mb} and the percent weight of aggregate as

$$\text{VMA} = \left(100 - G_{mb}\frac{P_s}{G_{sb}}\right) \tag{8.9}$$

The percent of the voids filled with asphalt, %VFA, is determined as

$$\text{VFA} = 100\left(\frac{\text{VMA} - \text{VTM}}{\text{VMA}}\right) \tag{8.10}$$

Sample Problem 8.2

An asphalt concrete specimen has the following properties:

- asphalt content = 5.9% by total weight of mix
- bulk specific gravity of the mix = 2.457
- theoretical maximum specific gravity = 2.598
- bulk specific gravity of aggregate = 2.692

Calculate the percents VTM, VMA, and VFA.

Solution

$$\text{VTM} = 100\left(1 - \frac{G_{mb}}{G_{mm}}\right) = 100\left(1 - \frac{2.457}{2.598}\right) = 5.4\%$$

$$\text{VMA} = \left(100 - G_{mb}\frac{P_s}{G_{sb}}\right) = \left(100 - 2.457\frac{100 - 5.9}{2.692}\right) = 14.1\%$$

$$\text{VFA} = 100\left(\frac{\text{VMA} - \text{VTM}}{\text{VMA}}\right) = 100\left(\frac{14.1 - 5.4}{14.1}\right) = 61.7\%$$

8.7.3 Superpave Mix Design

The Superpave mix-design process consists of

- ◆ Selection of aggregates
- ◆ Selection of binder
- ◆ Determination of the design aggregate structure
- ◆ Determination of the design binder content
- ◆ Evaluation of moisture susceptibility

Aggregate Selection Aggregate properties under the Superpave mix-design system are described as either consensus or source properties. The following consensus aggregate properties are required:

- ◆ coarse aggregate angularity measured by the percentage of fractured faces
- ◆ fine aggregate angularity (AASHTO TP 33) (see apparatus in Figure 4.3)
- ◆ flat and elongated particles (ASTM D4791)
- ◆ clay content (ASTM D2419)

Specification limits for these properties depend on the traffic level and how deep under the pavement surface the materials will be used, as shown in Table 8.2. In addition to these properties, highway agencies may consider other factors that are critical to the specific local conditions. These are called source properties and may include Los Angeles abrasion (see apparatus in Figure 4.4), soundness, and deleterious materials. Source properties are defined at the local level; consensus properties are defined at the national level.

Aggregate used in asphalt concrete must be well graded. Superpave recommends using the 0.45 power chart discussed in Chapter 4. The gradation curve should go through control points specified by Superpave. Figure 8.11 shows the gradation requirements for the 12.5 mm nominal-sized mix.

To control segregation, aggregates are sorted into stockpiles based on size. The designer of an asphalt concrete mix must select a blend of stockpiles that meets the source, consensus, and gradation requirements.

Binder Selection The binder is selected based on the maximum and minimum pavement temperatures, as discussed earlier. In addition to the specification tests, the specific gravity and the rotational viscosity versus temperature relationship for the selected asphalt binder must be measured. The specific gravity is needed for the void analysis. The viscosity-temperature relationship is needed to determine the required mixing and compaction temperatures. The Super-

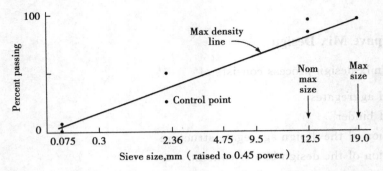

Figure 8.11 Superpave gradation limits for 12.5 mm nominal maximum size.

pave method requires mixing the asphalt and aggregates at a temperature at which the rotational viscosity of the asphalt binder is 0.170 ± 20 Pa·s and the compacting temperature corresponds to a viscosity of 0.280 ± 30 Pa·s.

Table 8.2 Superpave Consensus Aggregate Properties

Design Level	Course Aggregate Angularity (%min)	Fine Aggregate Angularity (%min)	Sand Equivalency (%min)	Flat and Elongated (%min)
Light Traffic	55/—	—	40	—
Medium Traffic	75/—	40	40	10
Heavy Traffic	85/80*	45	45	10

* 85/80 denotes minimum percentages of one fractured face/two or more fractured faces.

Design Aggregate Structure After selecting the appropriate aggregate, binder, and modifiers (if any), trial specimens are prepared with three different aggregate gradations and asphalt contents. There are equations for estimating the optimum asphalt content for use in the specimens prepared for determining the design aggregate structure. However, these equations are empirical and the designer is given latitude to estimate the asphalt content. Specimens are compacted using the Superpave gyratory compactor (Figure 8.7) with a gyration angle of 1.25 degrees and a constant vertical pressure of 600 kPa. The number of gyrations used for compaction is determined based on the traffic level, as shown in Table 8.3.

Table 8.3 Number of Gyrations at Specific Design Traffic Levels

	Traffic Level (10^6 ESAL*)			
	<0.3	0.3 to 3	3 to 30	>30
N_{ini}	6	7	8	9
N_{des}	50	75	100	125
N_{max}	75	100	125	205

* ESAL is the 18,000-lb equivalent single axle load. It is a design factor used in the design of pavement that considers both traffic volume and loads (Huang 2004).

As shown in Table 8.3, the Superpave method recognizes three critical stages of compaction, initial, design, and maximum. The design compaction level N_{des} corresponds to the compaction that is anticipated at the completion of the construction process. The maximum compaction N_{max} corresponds to the ultimate density level that will occur after the pavement has been subjected to traffic for a number of years. The initial compaction level N_{ini} was implemented to assist with identifying "tender" mixes. A tender mix lacks stability during construction, and hence will displace under the rollers rather than densifying. For the initial stage of determining the design aggregate structure, samples are compacted with N_{des} gyrations. The volumetric properties are determined by measuring the bulk specific gravity G_{mb} of the compacted mix and the maximum theoretical specific gravity G_{mm} of a loose mix with the same asphalt content and aggregate composition. The volumetric parameters, VTM, VMA, and VFA, are determined and checked against the criteria. Two additional parameters are evaluated in the Superpave method: percent of G_{mm} at N_{ini} and the dust-to-effective binder content ratio. The percent of G_{mm} at N_{ini} is determined as

$$\text{Percent } G_{mm}, N_{ini} = \text{Percent } G_{mm}, N_{des} \frac{h_{des}}{h_{ini}} \tag{8.11}$$

where h_{ini} and h_{des} are the heights of the specimen at the initial and design number of gyrations, respectively. Note that the percent G_{mm}, N_{des} is equal to $(100 - \text{VTM})$.

The dust-to-effective binder ratio is the percent of aggregate passing the 0.075 mm (#200) sieve divided by the effective asphalt content, computed as

$$P_{ba} = 100 \left(\frac{G_{se} - G_{sb}}{G_{sb} G_{se}} \right) G_b \tag{8.12}$$

$$P_{be} = P_b - \left(\frac{P_{ba}}{100} \right) P_s \tag{8.13}$$

$$D/B = \frac{P_D}{P_{be}} \tag{8.14}$$

where

D/B = dust to binder ratio
P_{ba} = percent absorbed binder based on the mass of aggregates
G_{sb} = bulk specific gravity of aggregate
P_D = percent dust, or % of aggregate passing the 0.075 mm sieve
P_{be} = percent effective binder content

The Superpave method requires determining the volumetric properties at 4% VTM. The samples prepared for the evaluation of the design aggregate content are not necessarily at the binder content required for achieving this level of air voids. Therefore, the results of the volumetric evaluation are "corrected" to four percent air voids, as

$$P_{b,est} = P_{bt} - 0.4(4 - VTM_t) \tag{8.15}$$

$$VMA_{est} = VMA_t + C(4 - VTM_t) \tag{8.16}$$

$$C = 0.1 \text{ for } VTM_t < 4.0\%$$

$$C = 0.2 \text{ for } VTM_t \geq 4.0\%$$

$$VFA_{est} = 100 \frac{VMA_t - 4.0}{VMA_t} \tag{8.17}$$

$$P_{be,est} = P_{b,est} - \frac{P_s G_b (G_{se} - G_{sb})}{G_{se} G_{sb}} \tag{8.18}$$

$$D/B_{est} = \frac{P_D}{P_{be,est}} \tag{8.19}$$

where

$P_{b,est}$ = adjusted estimated binder content

VMA_{est} = adjusted VMA

VMA_t = VMA determined from volumetric analysis

VFA_{est} = adjusted VFA

VTM_t = VTM determined from the volumetric analysis

$P_{be,est}$ = adjusted percent effective binder

D/B_{est} = adjusted dust-to-binder ratio

P_D = percent aggregate passing 0.075 mm sieve

In Equation 8.17, the air voids are assumed to be 4.0%, the target air void content for a Superpave mix design.

The adjusted results of the design aggregate structure evaluation are compared to the Superpave mix-design criteria, as shown in Table 8.4.

The design aggregate blend is the one whose adjusted volumetric parameters meet all of the criteria. In the event that more than one of the blends meets all of the criteria, the designer can use discretion for the selection of the blend. A pair of samples is compacted at the adjusted binder content for the selected aggregate blend. The average percent of maximum theoretical specific gravity is determined and compared to the design criteria. If successful, the procedure continues with the determination of the design binder content. If unsuccessful, the design process is started over with the selection of another blend of aggregates.

Table 8.4 Superpave Mix Design Criteria

Design Air Voids						4%
Dust to Effective Asphalt[1]						0.6~1.2
Tensile strength ratio						80% min

	Nominal Maximum Size(mm)					
	37.5	25	19	12.5	9.5	4.75
Minimum VMA(%)	11	12	13	14	15	16

	G_{mm} and VFA Requirements			
Design EASL in millions	Percent Maximum Theoretical Specific Gravity			Percent Voids Filled with Asphalt[2,3,4]
	N_{ini}	N_{des}	N_{max}	
<0.3	≤91.5	96	≤98.0	70~80
0.3~3	≤90.5	96	≤98.0	65~78
3~10	≤89.0	96	≤98.0	65~75
10~30	≤89.0	96	≤98.0	65~75
≥30	≤89.0	96	≤98.0	65~75

Notes: 1. Dust-to-binder ratio range is 0.9 to 2.0 for 4.75 mm mixes.
2. For 9.5 mm nominal maximum aggregate size mixes and design VFA≥3 million, VFA range is 73% to 76% and for 4.75 mm mixes the range is 75% to 78%.
3. For 25 mm nominal maximum aggregate size mixes, the lower limit of the VFA range shall be 67% for design traffic levels <0.3 million ESALS.
4. For 37.5 mm nominal maximum aggregate size mixes, the lower limit of the VFA range shall be 64% for all design traffic levels.

Sample Problem 8.3

Select a 19-mm Superpave design aggregate structure based on the following data:

Data	Blend		
	1	2	3
G_{mb}	2.457	2.441	2.477
G_{mm}	2.598	2.558	2.664
G_b	1.025	1.025	1.025
$P_b(\%)$	5.9	5.7	6.2
$P_s(\%)$	94.1	94.3	93.8
$P_d(\%)$	4.5	4.5	4.5
G_{sb}	2.692	2.688	2.665
h_{ini} (mm)	125	131	125
h_{des} (mm)	115	118	115

Solution

In this problem, the lab data were entered into an Excel spreadsheet and the appropriate equations were entered, producing the results shown in the accompanying table. The steps in the calculation are shown for the first blend only; the reader can verify the calculations for the other blends.

Volumetric Analysis				Blend			
Computed	Equation		Using Data for Blend 1	1	2	3	
G_{se}	9.6	$\dfrac{P_s}{\left(\dfrac{100}{G_{mm}} - \dfrac{P_b}{G_b}\right)}$	$\dfrac{94.1}{\left(\dfrac{100}{2.598} - \dfrac{5.9}{1.025}\right)}$	2.875	2.812	2.979	
VTM(%)	9.8	$100\left(1 - \dfrac{G_{mb}}{G_{mm}}\right)$	$100\left(1 - \dfrac{2.457}{2.598}\right)$	5.4	4.6	7.0	
VMA(%)	9.9	$\left(100 - G_{mb}\dfrac{P_s}{G_{sb}}\right)$	$\left(100 - 2.457\dfrac{94.1}{2.692}\right)$	14.1	14.4	12.8	
VFA(%)	9.10	$100\left(\dfrac{VMA - VTM}{VMA}\right)$	$100\left(\dfrac{14.1 - 5.4}{14.1}\right)$	61.7	68.1	45.3	
%G_{mm}, N_{ini}	9.11	Percent G_{mm}, $N_{des}\dfrac{h_{des}}{h_{ini}}$	$94.6\dfrac{115}{125}$	87.0	86.0	85.5	
P_{ba}(%)	9.12	$100\left(\dfrac{G_{se} - G_{sb}}{G_{sb}G_{se}}\right)G_b$	$100\left(\dfrac{2.875 - 2.692}{2.692 * 2.875}\right)1.025$	2.42	1.68	4.05	
P_{be}(%)	9.13	$P_b - \left(\dfrac{P_{ba}}{100}\right)P_s$	$5.9 - \left(\dfrac{2.42}{100}\right)94.1$	3.62	4.12	2.4	
D/B	9.14	$\dfrac{P_D}{P_{be}}$	$\dfrac{4.5}{3.62}$	1.2	1.1	1.9	
Adjusted Values							Criteria
$P_{b,est}$(%)	9.15	$P_{bt} - 0.4(4 - VTM_t)$	$5.9 - 0.4(4 - 5.4)$	6.5	5.9	7.4	
VMA_{est}(%)	9.16	$VMA_t + C(4 - VTM_t)$	$14.1 + 0.2(4 - 5.4)$	13.8	14.3	12.2	≥ 13
VFA_{est}(%)	9.17	$100\dfrac{VMA_t - 4.0}{VMA_t}$	$100\dfrac{13.8 - 5.4}{13.8}$	71.0	72.0	67.2	65—75
$P_{be,est}$(%)	9.18	$P_{b,est} - \dfrac{P_s G_b (G_{se} - G_{sb})}{G_{se} G_{sb}}$	$6.5 - \dfrac{94.1 * 1.025(2.875 - 2.692)}{2.875 * 2.692}$	4.2	4.3	3.6	
D/B_{est}	9.19	$\dfrac{P_D}{P_{be,est}}$	$\dfrac{4.5}{4.2}$	1.1	1.0	1.3	0.8—1.2

Although both mixes 1 and 2 meet the criteria, blend 2 is preferred, since it has a higher VMA value.

Design Binder Content The design binder content is obtained by preparing eight specimens—two replicates at each of four binder contents: estimated optimum binder content, 0.5% less than the optimum, 0.5% more than the optimum, and 1% more than the optimum. Specimens are compacted using N_{des} gyrations. The volumetric properties are computed and plots are prepared of each volumetric parameter versus the binder content. The binder content that corresponds to 4% VTM is determined. Then the plots are used to determine the volumetric parameters at the selected binder content. If the properties meet the criteria in Table 8.4, two specimens are prepared and compacted using N_{max} gyrations. If these specimens meet the criteria, then the moisture sensitivity of the mix is determined.

Sample Problem 8.4

Based on the previous problem, determine the recommended asphalt content according to Superpave mix design for an equivalent single axle load(ESAL)of 20 millions.

Solution

Samples were prepared at four asphalt contents. An Excel spreadsheet was prepared to present and analyze the data. The following results are obtained:

The results are plotted against binder content, as shown in Figure SP8.2.

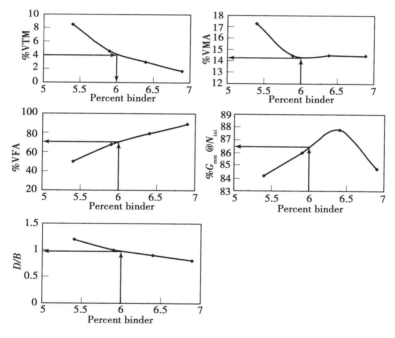

Figure SP8.2

Design Binder Content

Data	Asphalt Content Trial			
	1	2	3	4
$P_b(\%)$	5.4	5.9	6.4	6.9
G_{mb}	2.351	2.441	2.455	2.469
G_{mm}	2.570	2.558	2.530	2.510

Continued

Data	Asphalt Content Trial			
	1	2	3	4
G_b	1.025	1.025	1.025	1.025
$P_s(\%)$	94.6	94.1	93.6	93.1
$P_d(\%)$	4.5	4.5	4.5	4.5
G_{sb}	2.688	2.688	2.688	2.688
h_{ini} (mm)	125	131	126	130
h_{des} (mm)	115	118	114	112

Design Binder Content

Volumetric Analysis

Computed	Equation	Asphalt Content Trial			
		1	2	3	4
G_{se}	9.6	2.812	2.812	2.812	2.812
VTM(%)	9.8	8.5	4.6	3.0	1.6
VMA(%)	9.9	17.3	14.5	14.5	14.5
VFA(%)	9.10	50.9	68.3	79.3	89.0
%G_{mm}, N_{ini}	9.11	84.2	86.0	87.8	84.7
$P_{ba}(\%)$	9.12	1.68	1.68	1.68	1.68
$P_{be}(\%)$	9.13	3.81	4.32	4.83	5.34
D/B	9.14	1.2	1.0	0.9	0.8

The figure shows that, at 4% VTM, the binder content is 6%. The design values at 6% binder content are

- VMA=14.5%
- VFA=71%
- G_{mm} @ N_{ini}=86.5%
- D/B ratio=0.9

These values satisfy the design criteria shown in Table 8.4. Therefore, the design binder content is 6.0%.

Moisture Sensitivity Evaluation The moisture sensitivity of the design mixture is determined using the AASHTO T283 procedure on six specimens prepared at the design binder content and 7% air voids. Three specimens are conditioned by vacuum saturation, then freezing and thawing; three other specimens are not conditioned. The tensile strength of each sample is measured using a Marshall stability machine with a modified loading head. The tensile strength ratio is determined as the ratio of the average tensile strength of conditioned specimens to that of unconditioned specimens. The minimum Superpave criterion for tensile strength ratio is 80%.

8.7.4 Superpave Simple Performance Tests(SPT)

The volumetric procedure has been widely implemented with Successful results. However, the method lacks a strength test to verify the suitability of the Superpave mixes. Research completed in NCHRP Project 9-19 has identified three candidate simple performance tests(SPT) for HMA(Witczak et al. 2002). Tests based on measurement of dynamic modulus(for both of permanent deformation and fatigue cracking), flow time(permanent deformation), and flow number(permanent deformation) were selected for further field validation. The three tests used to obtain these parameters are the dynamic modulus test, triaxial static creep test, and triaxial repeated load permanent deformation test. All these tests use cylindrical specimens 100 mm in diameter and 150 mm high. Specimens are cored in the lab from specimens compacted using the Superpave gyratory compactor(Figure 8.7). Figure 8.12 shows a photo of the SPT.

Figure 8.12 Superpave simple performance test assembly.

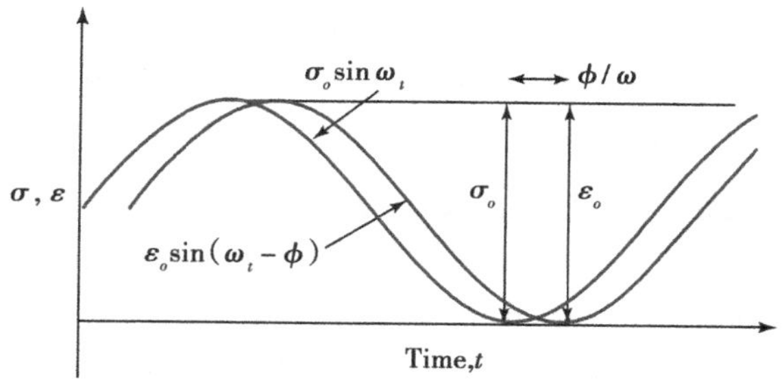

Figure 8.13 Stress and strain pulses for the dynamic modulus test.

Dynamic Modulus Test　The dynamic modulus test in triaxial compression has been around the pavement community for many years (ASTM D3497). The test consists of applying an axial sinusoidal compressive stress to an unconfined or confined HMA cylindrical test specimen, as shown in Figure 8.13.

Assuming that HMA is a linear viscoelastic material, a complex number called the complex modulus, E^*, can be obtained from the test to define the relationship between stress and strain. The absolute value of the complex modulus, $|E^*|$, is called the dynamic modulus. The dynamic modulus is mathematically defined as the peak dynamic stress σ_o divided by the peak recoverable axial strain ε_o:

$$|E^*| = \frac{\sigma_o}{\varepsilon_o} \tag{8.20}$$

The real and imaginary portions of the complex modulus E^* can be written as

$$E^* = E' + iE'' \tag{8.21}$$

E' is generally referred to as the storage or elastic modulus component of the complex modulus and E'' is referred to as the loss or viscous modulus. The phase angle ϕ is the angle by which ε_o lags behind σ_o. It is an indicator of the viscous properties of the material being evaluated. Mathematically, this is expressed as

$$E^* = |E^*|\cos\phi + i|E^*|\sin\phi \tag{8.22}$$

where

$\phi = \dfrac{t_i}{t_p} \times 360$

t_i = time lag between of stress and strain, s

t_p = time for a stress cycle, s

i = imaginary number

For a pure elastic material, $\phi = 0$, and the complex modulus E^* is equal to the absolute value, or the dynamic modulus. For a pure viscous material, $\phi = 90°$.

The dynamic modulus obtained from this test is indicative of the stiffness of the asphalt mixture at the selected temperature and load frequency. The dynamic modulus is correlated to both rutting and fatigue cracking of HMA.

Triaxial Static Creep Test In the static compressive creep test, a total strain-time relationship for a mixture is measured in the laboratory under unconfined or confined conditions. The static creep test, using either one load-unload cycle or incremental load-unload cycles, provides sufficient information to determine the instantaneous elastic (recoverable) and plastic (irrecoverable) components (time independent), and the viscoelastic and viscoplastic components (time dependent) of the material's response. In this test, the compliance, $D(t)$ is determined by dividing the strain as a function of time $\varepsilon(t)$, by the applied constant stress σ_o:

$$D(t) = \frac{\varepsilon(t)}{\sigma_o} \tag{8.23}$$

Figure 8.14 shows typical test results between the calculated compliance and loading time. As shown, the compliance can be divided into three major zones: primary zone, secondary zone, and tertiary flow zone. The time at which tertiary flow starts is referred to as the *Flow Time*. The flow time is a significant parameter in evaluating the rutting resistance of HMA.

Triaxial Repeated Load Permanent Deformation Test Another approach to measuring the permanent deformation characteristics of HMA is to use a repeated load test for several thousand repetitions and to record the cumulative permanent deformation as a function of the number of load repetitions. In this test, a haversine pulse load consisting of a 0.1 second and 0.9 second dwell (rest) time is applied for the test duration, typically about three hours or 10,000 loading cycles.

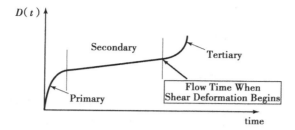

Figure 8.14 Compliance versus loading time for the static triaxial creep test.

Results from the repeated load tests typically are presented in terms of the cumulative permanent strain versus the number of loading cycles. Figure 8.15 illustrates a typical relationship between the cumulative plastic strain and number of load cycles. In a manner similar to the triaxial static creep test, the cumulative permanent strain curve can be divided into three zones: primary, secondary, and tertiary. The cycle number at which tertiary flow starts is referred to as the Flow Number. In addition to the flow number, the test can provide the resilient strain and modulus, all of which are correlated to the rutting resistance of HMA.

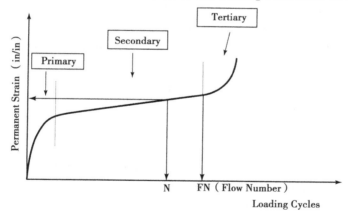

Figure 8.15 Typical relationship between total cumulative plastic strain and loading cycles.

8.7.5 Marshall Method of Mix Design

The basic steps required for performing Marshall mix design are as follows (The Asphalt Institute 1995):

1. aggregate evaluation
2. asphalt cement evaluation

3. specimen preparation
4. Marshall stability and flow measurement
5. density and voids analysis
6. design asphalt content determination

1. **Aggregate Evaluation** The aggregate characteristics that must be evaluated before it can be used for an asphalt concrete mix include the durability, soundness, presence of deleterious substances, polishing, shape, and texture. Agency specifications define the allowable ranges for aggregate gradation. The Marshall method is applicable to densely graded aggregates with a maximum size of not more than 25 mm.

2. **Asphalt Cement Evaluation** The grade of asphalt cement is selected based on the expected temperature range and traffic conditions. Most highway agencies have specifications that prescribe the grade of asphalt for the design conditions.

3. **Specimen Preparation** The full Marshall mix-design procedure requires 18 specimens 101.6 mm in diameter and 63.5 mm high. The stability and flow are measured for 15 specimens. In addition, 3 specimens are used to determine the theoretical maximum specific gravity G_{mm}. This value is needed for the void and density analysis. The specimens for the theoretical maximum specific gravity determination are prepared at the estimated design asphalt content. Samples are also required for each of five different asphalt contents; the expected design asphalt content, $\pm 0.5\%$ and $\pm 1.0\%$. Engineers use experience and judgment to estimate the design asphalt content.

   Specimen preparation for the Marshall method uses the Marshall compactor discussed earlier (Figure 8.8). The Marshall method requires mixing of the asphalt and aggregates at a temperature where the kinematic viscosity of the asphalt cement is 170 ± 20 cSt and compacting temperature corresponds to a viscosity of 280 ± 30 cSt.

   The Asphalt Institute permits three different levels of energy to be used for the preparation of the specimens: 35, 50, and 75 blows on each side of the sample. Most mix designs for heavy-duty pavements use 75 blows, since this better simulates the required density for pavement construction.

4. **Marshall Stability and Flow Measurement** The Marshall stability of the asphalt concrete is the maximum load the material can carry when tested in the Marshall apparatus, Figure 8.16. The test is performed at a deformation rate of 51 mm/min. (2 in./min.) and a temperature of 60 ℃. The Marshall flow is the deformation of the specimen when the load starts to decrease. Stability is reported in newtons (pounds) and flow is reported in units of 0.25 mm of deformation. The stability of specimens that are not 63.5 mm thick is adjusted by

multiplying by the factors shown in Table 8.5. All specimens are tested and the average stability and flow are determined for each asphalt content.

Figure 8.16 Marshall stability machine.

Table 8.5 Marshall Stability Adjustment Factors

Approximate Thickness of Specimen, mm(in.)	Adjustment Factor	Approximate Thickness of Specimen, mm(in.)	Adjustment Factor
50.8(2)	1.47	65.1(2 9/16)	0.96
52.4(2 1/16)	1.39	66.7(2 5/8)	0.93
54.0(2 1/8)	1.32	68.3(2 11/16)	0.89
55.6(2 3/16)	1.25	69.8(2 3/4)	0.86
57.2(2 1/4)	1.19	71.4(2 13/16)	0.83
58.7(2 5/16)	1.14	73.0(2 7/8)	0.81
60.3(2 3/8)	1.09	74.6(2 15/16)	0.78
61.9(2 7/16)	1.04	76.2(3)	0.76
63.5(2 1/2)	1.00		

5. **Density and Voids Analysis**   The values of VTM, VMA, and VFA are determined as using Equations 8.8, 8.9, and 8.10, respectively.

6. **Design Asphalt Content Determination**   Traditionally, test results and calculations are tabulated and graphed to help determine the factors that must be used in choosing the optimum

Table 8.6 Examples of Mix Design Measurements and Calculations by the Marshall Method (The Asphalt Institute, 1995)

% AC by Wt. of Mix, Spec. No.	Spec. Height, mm	Wt. in Air, g	Wt. in Water, g	SSD Wt., g	Bulk Vol., cm³	Bulk Sp. Gr.	Max. Theo. Sp. Gr. (loose mix)	% Air Voids	% VMA	% VFA	Measured Stability, kN	Adjusted Stability, kN	Flow, 0.25 mm
3.5—A		1240.6	726.4	1246.3	519.9	2.386					10.9	10.9	8
3.5—B		1238.7	723.3	1242.6	519.3	2.385					10.8	10.8	7
3.5—C		1240.1	724.1	1245.9	521.8	2.377					11.2	11.2	7
Average						2.383	2.570	7.3	14.0	48.0		10.9	7
4.0—A		1244.3	727.2	1246.6	519.4	2.396					9.7	9.7	9
4.0—B		1244.6	727.0	1247.6	520.6	2.391					10.1	10.1	9
4.0—C		1242.6	727.9	1244.0	516.1	2.408					10.3	10.3	8
Average						2.398	2.550	6.0	13.9	57.1		10.1	9
4.5—A		1249.3	735.8	1250.2	414.4	2.429					10.8	10.8	9
4.5—B		1250.8	728.1	1251.6	523.5	2.389					10.7	10.3	9
4.5—C		1251.6	735.3	1253.1	517.8	2.417					10.4	10.4	9
Average						2.412	2.531	4.7	13.9	66.1		10.5	9
5.0—A		1256.7	739.8	1257.6	517.8	2.427					10.2	10.2	9
5.0—B		1258.7	742.7	1259.3	516.6	2.437					9.7	9.7	8
5.0—C		1258.4	737.5	1259.1	521.6	2.413					10.0	10.0	9
Average						2.425	2.511	3.4	13.8	75.2		10.0	9
5.5—A		1263.8	742.6	1264.3	521.7	2.422					9.8	9.8	9
5.5—B		1258.8	741.4	1259.4	518.0	2.430					10.2	10.2	10
5.5—C		1259.0	742.5	1259.5	517.0	2.435					9.8	10.0	9
Average						2.429	2.493	2.5	14.1	82.1		10.0	9

Notes: AC-20 binder; $G_b = 1.030$, $G_{sb} = 2.674$, Absorbed AC of aggregate; 0.6%, $G_{se} = 2.717$, Compaction; 75 blows

asphalt content. Table 8.6 presents examples of mix design measurements and calculations. Figure 8.17 shows plots of results obtained from Table 8.6, which include asphalt content versus air voids, VMA, VFA, unit weight, Marshall stability, and Marshall flow.

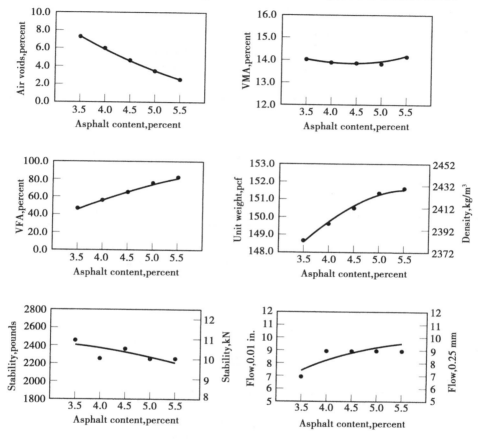

Figure 8.17 Graphs used for Marshall mix design analysis. (See Table 8.6) (The Asphalt Institute 1995).

The design asphalt content is usually the most economical one that will satisfactorily meet all of the established criteria. Different criteria are used by different agencies. Table 8.7 and 8.8 depict the mix design criteria recommended by The Asphalt Institute. Figure 8.18 shows an example of the narrow range of acceptable asphalt contents. The asphalt content selection can be adjusted within this narrow range to achieve a mix that satisfies the requirements of a specific project. Other agencies, such as the National Asphalt Paving Association, use the asphalt cement content at 4% air voids as the design value, and then check that the other factors meet the criteria. If the Marshall stability, Marshall flow, VMA, or VFA fall outside the allowable range, the mix must be redesigned using an adjusted aggregate gradation or new material

234   Materials for Civil and Construction Engineers

sources.

The laboratory-developed mixture design forms the basis for the initial job mix formula (JMF). The initial JMF should be adjusted to reflect the slight differences between the laboratory-supplied aggregates and those used in the field.

Table 8.7 Asphalt Institute Criteria for Marshall Mix Design (The Asphalt Institute, 1995)

	Traffic Level					
	Light		Medium		Heavy	
Compaction (blows)	35		50		75	
	Minimum	Maximum	Minimum	Maximum	Minimum	Maximum
Stability, kN	3.34	—	5.34	—	8.01	—
Flow, 0.25 mm	8	18	8	16	8	14
Air Voids, %	3	5	3	5	3	5
VMA, %	Use the criteria in Table 9.14					
VFA, %	70	80	65	78	65	75

Table 8.8 Minimum Percent Voids in Mineral Aggregate (VMA) (The Asphalt Institute, 1995)

Nominal Maximum Particle Size[1]	Minimum VMA, Percent		
	Design Air Voids[2]		
	3.0	4.0	5.0
2.36 mm (No. 8)	19.0	20.0	21.0
4.75 mm (No. 4)	16.0	17.0	18.0
9.5 mm (3/8 in.)	14.0	15.0	16.0
12.5 mm (1/2 in.)	13.0	14.0	15.0
19.0 mm (3/4 in.)	12.0	13.0	14.0
25.0 mm (1.0 in.)	11.0	12.0	13.0

1. The nominal maximum particle size is one size larger than the first sieve to retain more than 10 percent.
2. Interpolate minimum VMA for design air void values between those listed.

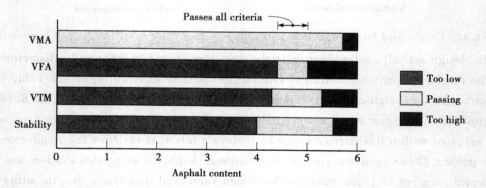

Figure 8.18 An example of the narrow range of acceptable asphalt contents. (The Asphalt Institute 1995).

Sample Problem 8.5

The Marshall method was used to design an asphalt concrete mixture. An AC-30 asphalt cement with a specific gravity(G_b) of 1.031 was used. The mixture contains a 9.5 mm nominal maximum particle size aggregate with a bulk specific gravity(G_{sb}) of 2.696. The theoretical maximum specific gravity of the mix(G_{mm}) at asphalt content of 5.0 % is 2.470. Trial mixes were made with average results as shown in the following table:

Asphalt Content(P_b) (% by Weight of Mix)	Bulk Specific Gravity (G_{mb})	Corrected Stability(kN)	Flow(0.25 mm)
4.0	2.360	6.3	9
4.5	2.378	6.7	10
5.0	2.395	5.4	12
5.5	2.405	5.1	15
6.0	2.415	4.7	22

Determine the design asphalt content using me Asphalt Institute design criteria for medium traffic. (Table 8.7). Assume a design air void content of 4% when using Table 8.8.

Solution

Analysis steps:

1. Determine the effective specific gravity of the aggregates G_{se}, using Equation 8.6:

$$G_{se} = \frac{P_s}{\left(\frac{100}{G_{mm}} - \frac{P_b}{G_b}\right)} = \frac{100 - 5.0}{\left(\frac{100}{2.470} - \frac{5.0}{1.031}\right)} = 2.666$$

The calculations in steps 2—5 are for 4.0% asphalt content as an example. Repeat for other asphalt contents.

2. Use G_{se} to determine G_{mm} for the other asphalt contents, using Equation 8.5:

$$G_{mm} = \frac{100}{\left(\frac{P_s}{G_{se}} + \frac{P_b}{G_b}\right)} = \frac{100}{\left(\frac{100-4.0}{2.666} + \frac{4.0}{1.031}\right)} = 2.507$$

3. Compute voids in the total mix for each asphalt content, using Equation 8.8:

$$\text{VTM} = 100\left(1 - \frac{G_{mb}}{G_{mm}}\right) = 100\left(1 - \frac{2.360}{2.507}\right) = 5.9$$

4. Compute voids in mineral aggregate, using Equation 8.9:

$$\text{VMA} = 100 - \left(\frac{G_{mb} P_s}{G_{mm}}\right) = 100 - \left(2.360 \times \frac{100-4.0}{2.696}\right) = 16.0$$

5. Compute voids filled with asphalt, using Equation 8.10:

$$\text{VFA} = 100 \frac{(\text{VMA} - \text{VTM})}{\text{VMA}} = 100 \frac{(16.0 - 5.9)}{16.0} = 63.3$$

6. A summary of all calculations is given in the following table:

P_b(%)	G_{mb}	Corrected Stability(kN)	Flow, (0.25 mm)	G_{mm}	G_{se}	VTM(%)	VMA(%)	VFA(%)
4.0	2.360	6.3	9	2.507		5.9	16.0	63.3
4.5	2.378	6.7	10	2.488		4.4	15.8	71.9
5.0	2.395	5.4	12	2.470	2.666	3.0	15.6	80.5
5.5	2.405	5.1	15	2.452		1.9	15.7	87.8
6.0	2.415	4.7	22	2.434		0.8	15.8	95.0

7. Plot stability, flow and volumetric parameters versus P_b. (See Figure SP8.5.)
8. Determine the asphalt content that corresponds to VTM=4% and the corresponding parameters. Compare with criteria:

	From Graphs	Criteria
P_b@4%	4.6	
Stability(kN)	6.6	5.34(min)
Flow(0.25 mm)	10.5	8 to 16
G_{mb}	2.383	NA
VMA(%)	15.7	15.0(min)
VFA(%)	75	65 to 78

Therefore, the design asphalt content is 4.6%.

8.7.6 Hveem Method of Mix Design

The basic steps required for performing Hveem mix design are the following (The Asphalt Institute 1995):

1. aggregate evaluation
2. asphalt cement evaluation
3. evaluation of centrifuge kerosene equivalent of fine aggregate
4. evaluation of surface capacity of coarse aggregate
5. estimation of optimum asphalt content
6. specimen preparation
7. measurement of the Hveem stability
8. density and voids analysis
9. determination of design asphalt content.

The evaluation of aggregate and asphalt cement is performed as in the Marshall method of mix design. The Hveem method requires measuring aggregate properties and using a series of

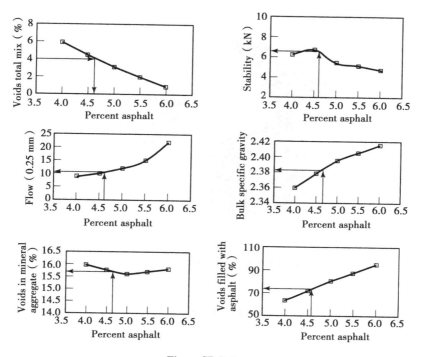

Figure SP 8.5

charts to estimate the design asphalt content (The Asphalt Institute 1995).

Three cylindrical specimens 102 mm in diameter and 63.5 mm high are prepared, using the California kneading compactor (Figure 8.9) according to ASTM D1561. Three asphalt contents near the estimated design value are used to fabricate the specimens. The Hveem stability of the specimens is determined using the Hveem stabilometer (Figure 8.19), according to ASTM D1560. The Hveem stabilometer is a device that allows for the application of a lateral pressure on the specimen while applying vertical load using a compression machine.

As in the Marshall method, the bulk specific gravity, theoretical maximum specific gravity, percent air voids (VTM), and density of all specimens are determined. The Hveem stability, density, and air voids are tabulated and plotted versus asphalt content. The optimum asphalt content for the design mix should be the highest asphalt content the mix will accommodate without reducing the stability or void content below the minimum values required by the design criteria.

The laboratory-developed mixture design forms the basis for the initial JMF. The initial JMF should be adjusted to consider the slight differences between the laboratory-supplied aggregates and those used in the field.

Figure 8.19 Hveem stabilometer.

8.7.7 Evaluation of Moisture Susceptibility

Since loss of bond between asphalt and aggregates (*stripping*) has become a significant form of asphalt pavement distress, several methods have been developed for evaluating the susceptibility of a mix to water damage. Most of the popular methods require the specimens to be at the optimum asphalt content and mix gradation.

The specimens are divided into two lots: reference specimens and conditioned specimens. A strength test is used to evaluate the strength before and after conditioning; the retained strength, the ratio of conditioned strength to reference strength, expressed in percent, is computed. Criteria are used to determine if the retained strength is adequate. The different techniques for evaluating moisture susceptibility vary, depending on the specimen preparation, conditioning procedures, and strength.

The immersion-compression test (ASTM D1075) has been used to evaluate moisture susceptibility. The method evaluates the retained compressive strength after vacuum saturation. Other methods use Marshall specimens, freezing and water soaking to condition the samples, and determining diametral strength and modulus values to evaluate the retained strength. Freezing the samples greatly increases the severity of the test.

There are several ways to alter asphalt concrete's susceptibility to water damage. Methods identified by the Asphalt Institute include the following:

1. increasing asphalt content
2. using a higher viscosity asphalt cement

3. cleaning aggregate of any dust and clay
4. adding antistripping additives
5. altering aggregate gradation

In addition, portland cement and lime have been used by some agencies as antistripping agents. Generally, when water damage susceptibility is a problem. the additive is added to the mix at three levels, and the water damage test is perfomed to determine the minimum amount of additive that can be used to increase the retained strength to an acceptable level. If an acceptable mix can be developed. Marshall or Hveem specimens are prepared, and the mix is tested to determine whether it meets the design criteria.

8.8 Characterization of Asphalt Concrete

Tests used to characterize asphalt concrete are somewhat different from those used to characterize other civil engineering materials, such as steel, portland cement concrete, and wood. One of the main reasons for this difference is that asphalt concrete is a nonlinear viscoelastic or viscoelastoplastic material. Thus, its response to loading is greatly affected by the rate of loading and temperature. Also, asphalt pavements are typically subjected to dynamic loads applied by traffic. Moreover, asphalt pavements do not normally fail due to sudden collapse under the effect of vehicular loads, but due to accumulation of permanent deformation in the wheel path (rutting), cracking due to repeated bending of the asphalt concrete layer (fatigue cracking), thermal cracking, excessive roughness of the pavement surface, migration of asphalt binder at the pavement surface (bleeding or flushing), loss of flexibility of asphalt binder due to aging and oxidation (raveling), loss of bond between the asphalt binder and aggregate particles due to moisture (stripping), or other factors. Therefore, most of the tests used to characterize asphalt concrete try to simulate actual field conditions.

Many laboratory tests have been used to evaluate asphalt concrete properties and to predict its performance in the field. These tests are performed on either laboratory-prepared specimens or cores taken from in-service pavements. These tests measure the response of the material to load, deformation, or environmental conditions, such as temperature, moisture, or freeze and thaw cycles. Some of these tests are based on empirical relations, while others evaluate fundamental properties. All tests on asphalt concrete are performed at accurately controlled test temperatures and rates of loading, since asphalt response is largely affected by these two parameters.

The Superpave tests used for mix design, as well as Marshall or Hveem tests discussed earlier, have been used to characterize asphalt concrete mixtures. Other tests are also being used, some of which are standardized by ASTM or AASHTO, while others have been used mostly for research. The next several sections discuss some of the common tests.

8.8.1 Indirect Tensile Strength

When traffic loads are applied on the pavement surface, tension is developed at the bottom of the asphalt concrete layer. Therefore, it is important to evaluate the tensile strength of asphalt concrete for the design of the layer thickness. In this test, a cylindrical specimen 102 mm in diameter and 64 mm high is typically used. A compressive vertical load is applied along the vertical diameter, using a loading device similar to that shown in Figure 8.20. The load is applied

Figure 8.20 Diametral loading for indirect tensile strength and resilient modulus tests.

by using two curved loading strips moving with a rate of deformation of 51 mm/min. Tensile stresses are developed in the horizontal direction, and when these stresses reach the tensile strength, the specimen fails in tension along the vertical diameter. The test is performed at a specified temperature. With 12.5 mm loading strips, the indirect tensile strength is computed as

$$\sigma_t = \frac{2P}{\pi t D} \qquad (8.24)$$

where

σ_t = tensile strength, MPa(psi)
P = load at failure, N(lb)
t = thickness of specimen, mm(in.)
D = diameter of specimen, mm(in.)

8.8.2 Diametral Tensile Resilient Modulus

To evaluate the structural response of the asphalt pavement system, the modulus of asphalt concrete material is needed. Since asphalt concrete is not a linear viscoelastic material, the modulus of elasticity, Young's modulus, is not applicable. The diametral tensile resilient modulus test (ASTM D4123) provides an analogous modulus, known as the resilient modulus. The test uses a cylindrical specimen 102 mm in diameter and 63.5 mm high. A pulsating load is applied along the vertical diameter, using a load guide device similar to that shown in Figure 8.20. The load is commonly applied with a duration of 0.1 second and a rest period of 0.9 second. After a few hundred repetitions, the recoverable horizontal deformation is measured using two linear variable differential transducers (LVDTs). Figure 8.21 shows typical load and horizontal de-

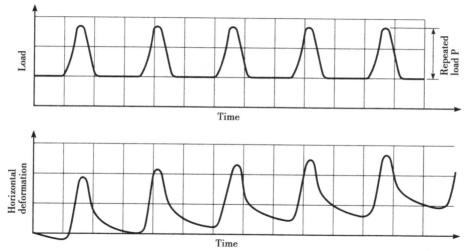

Figure 8.21 Typical load and horizontal deformation versus time during the resilient modulus test.

formation versus time relationships. Since the test is nondestructive, the test is repeated on the same specimen after rotating it 90°. The test is commonly performed at three temperatures: 5 ℃, 25 ℃, and 40 ℃. The diametral tensile resilient modulus is computed as

$$M_R = \frac{P(0.27 + v)}{t \cdot \Delta H} \tag{8.25}$$

where
- M_R = indirect tensile resilient modulus, MPa(psi)
- P = repeated load, N(lb)
- v = Poisson's ratio, typically 0.3, 0.35, and 0.4 at temperatures of 5 ℃, 25 ℃, and 40 ℃, respectively

t = thickness of specimen, mm(in.)

ΔH = sum of recoverable horizontal deformations on both sides of specimen, mm(in.)

Typical resilient modulus values of asphalt concrete are 6.89 GPa, 4.13 GPa, and 1.38 GPa at temperatures of 5 ℃, 25 ℃, and 40 ℃, respectively. The diametral tensile resilient modulus test is very sophisticated, because it measures very small deformations. Therefore, extreme caution must be exercised to align the specimen between the loading trips and to reduce possible rocking. Also, the load magnitude must be small enough to reduce the possibility of permanent deformation in the specimen, yet large enough to obtain measurable deformation.

Sample Problem 8.6

The resilient modulus test was performed on an asphalt concrete specimen and the following data were obtained:

diameter = 4.000 in.
thickness = 2.523 in.
repeated load = 559 lb
sum of recoverable horizontal deformations = 254×10^{-6} in.

Assuming a Poisson's ratio of 0.35, calculate the resilient modulus.

Solution

$$M_R = \frac{P(0.27+v)}{t \cdot \Delta H} = 559 \times (0.27+0.35)/(2.523 \times 254 \times 10^{-6})$$
$$= 541,000 \text{ psi}$$

8.8.3 Freeze and Thaw Test

The freeze and thaw test is performed to evaluate the effect of freeze and thaw cycles on the stiffness properties of asphalt concrete. Cylindrical specimens 102 mm in diameter and 64 mm high are used. Three specimens are tested for resilient modulus as discussed earlier, while the other three specimens are subjected to cycles of freeze and thaw, after which the resilient modulus is determined. The tensile strength ratio is computed by dividing the average resilient modulus of conditioned specimens by the average resilient modulus of unconditioned specimens, expressed in percent. A minimum tensile strength ratio is usually required to identify mixes that are not severely affected by freeze and thaw cycles.

8.8.4 Use of Rheological Models to Analyze Time-Dependent Response

Asphalt concrete is a viscoelastic material exhibiting a time-dependent response under load. Rheological models consisting of combinations of Hookean (spring) and Newtonean (dashpots) elements have been used to analyze the response of time-dependent materials, as discussed in Chapter 1. The Burgers model illustrated in Figure 1.12 can closely approximate the response of asphaltic mixtures (M amlouk 1984). Laboratory tests, such as the creep test, are used to obtain the parameters of the Burgers model using a curve-fitting procedure. Once these parameters are determined, the model can be used to predict the response of the material under different loading conditions. For example, Burgers model has been used to predict rutting of asphalt concrete pavement under the action of traffic loads.

8.9 Asphalt Concrete Production

Asphalt concrete is produced in either a batch plant or a continuous (drum) plant (The Asphalt Institute 1989). In the United States, batch plants were used extensively in the past; however, more energy efficient continuous plants are now preferred.

In continuous plants, aggregates of different gradations are placed in cold bins. The gradation proportions needed are taken from the cold bins by a cold feed elevator. Aggregates are transferred to the first part of the drum, where they are dried and heated. Hot asphalt cement is introduced in the last one-third of the drum; then aggregates and asphalt are mixed. Since asphalt concrete is produced continuously in this type of plant, it is transferred to a storage silo until placed in a truck and transported to the job site.

8.10 Recycling of Asphalt Concrete

Recycling pavement materials has a long history. However, recycling became more important in the mid-1970s, after the oil embargo, due to the increase in asphalt prices. In an effort to efficiently use available resources, there was a need to recycle or reuse old pavement materials (The Asphalt Institute 1989). Although the pavement could be badly deteriorated, the old asphalt concrete materials could be successfully reused in new pavements. Currently, recycling of old pavement materials is becoming a normal practice due to the following advantages:

1. economic saving of about 25% of the price of materials
2. energy saving in manufacturing and transporting raw materials
3. environmental saving by reducing the amount of required new materials and by eliminating the problem of discarding old materials
4. eliminating the problem of reconstruction of utility structures, curbs, and gutters associated with overlays
5. reducing the dead load on bridges due to overlays
6. maintaining the tunnel clearance, compared with overlays

Recycling, can be divided into three types: *surface recycling*, *central plant recycling*, and *inplace recycling*.

8.10.1 Surface Recycling

Surface recycling is defined as the reworking of the top 25 mm of the pavement surface using a heater-scarifier. The heater planing machine heats the pavement surface, which repairs minor cracks and roughness. Usually, a rejuvenating agent is added after heating, followed by slight scratching of the surface and compaction.

8.10.2 Central Plant Recycling

Central plant recycling is performed by milling the old pavement and sending the reclaimed asphalt pavement (RAP) to a central asphalt concrete plant, wthere it is mixed with some form of rejuvenating agent or soft asphalt and aggregates to produce hot-mixed asphalt concrete. If the RAP materials are mixed with the aggregates in a conventional asphalt concrete plant, they will burn and produce smoke, causing significant environmental problems. Therefore, the RAP materials are added to the pugmill (mixer) in the batch plant or added at midlength in the drum at the drum mix plant. The amount of recycled materials varies from 20% to 70%. The gradation of new aggregates is selected to correct any deficiency in the gradation of recycled aggregate. Typically, the grade of new asphalt cement is soft so that, when it is mixed with the old, hard asphalt, an appropriate consistency will result. The mix design of asphalt concrete, including recycled materials, is usually performed using either the Marshall or Hveem procedure.

In addition to hot-mix central plant recycling, cold central plant recycling can use new emulsified or cutback asphalt. However, the cold process will not have the quality of the hot-mixed material.

8.10.3 In-Place Recycling

In-place recycling is performed by ripping and pulverizing the old pavement surface and adding new aggregate, water, and asphalt emulsion. The old and new materials are mixed together in place, graded, and compacted. The surface is left to cure and is then used as a surface layer for low-volume roads. The recycled layer can also be used as a stabilized base, covered by an asphalt concrete surface.

8.11 Additives

Many types of additives (modifiers) are used to improve the properties of asphalt or to add special properties to the asphalt concrete mixtures (Roberts et al. 1996). Laboratory tests are usually performed and field performance is observed in order to evaluate the effect of the additives and to justify their cost. The effects of using additives should be carefully evaluated; otherwise premature pavement failure might result. The recyclability of modified asphalt mixtures is still being evaluated.

8.11.1 Fillers

Several types of fillers, such as crushed fines, portland cement, lime, fly ash, and carbon black can be added to asphalt concrete. Fillers are used to satisfy gradation requirements of materials passing the 0.075 mm (No. 200) sieve, to increase stability, to improve bond between aggregates and asphalt, or to fill the voids and thus reduce the required asphalt.

8.11.2 Extenders

Extenders such as sulfur and lignin are used to reduce the asphalt requirements, thus reducing the cost.

8.11.3 Rubber

Rubber has been used in asphalt concrete mixture in the form of natural rubber, styrene-butadiene (SBR), styrene-butadiene-styrene (SBS), or recycled tire rubber. Rubber increases elasticity and stiffness of the mix and increases the bond between asphalt and aggregates. Scrap rubber tires can be added to the asphalt cement(wet method)or added as crumb rubber to the aggregates(dry method).

8.11.4 Plastics

Plastics have been used to improve certain properties of asphalt. Plastics used include polyethylene, polypropylene, ethyl-vinyl-acetate(EVA), and polyvinyl chloride(PVC). They increase the stiffness of the mix, thus reducing the rutting potential. Plastics also may reduce the temperature susceptibility of asphalt and improve its performance at low temperatures.

8.11.5 Antistripping Agents

Antistripping agents are used to improve the bond between asphalt cement and aggregates, especially for water susceptible mixtures. Lime is the most commonly used antistripping agent and can be added as a filler or a lime slurry and mixed with the aggregates. Portland cement can be used as an alternative to lime.

8.11.6 Others

Other additives, such as fibers, oxidants, antioxidants, and hydrocarbons, have been used to modify certain asphalt properties' tensile strength and stiffness.

SUMMARY

Asphalt produced from crude oil is a primary road-building material. The civil engineer is directly involved with the specification and requirements for both the asphalt cement binder and the asphalt concrete mixtures. There are several methods for grading asphalt cements. The current trend is toward the use of the performance-grading method, used in the Superpave process developed through the Strategic Highway Research Program. This grading method directly ties the binder properties to pavement-performance parameters. Similarly, the Superpave mix design method uses performance tests to evaluate the mixture characteristics relative to expected field performance. This method will continue to be refined by highway agencies. With the support of the Federal Highway Administration, there is a concerted effort being placed on replacing the traditional Marshall and Hveem mix design methods with the Superpave procedures.

QUESTIONS AND PROBLEMS

8.1 What is the difference between tar and asphalt cement?

8.2 Discuss the main uses of asphalt.

8.3 Define what is meant by temperature susceptibility of asphalt. Discuss the effect on the performance of asphalt concrete pavements. Are soft asphalts used in hot or cold climates?

8.4 Temperature has a large effect on the asphalt viscosity. On one graph, plot the relationship between asphalt viscosity(logarithmic) and temperature for two cases: (a) a low-temperature susceptible asphalt and (b) a high-temperature susceptible asphalt. Label all axes and relations.

8.5 What is the significance of each one of these tests:
   a. flash point test
   b. RTFO procedure
   c. rotational viscometer test
   d. dynamic shear rheometer test
   e. penetration test

8.6 Discuss the aging that occurs in asphalt cement during mixing with aggregates and in service. How can the different types of aging of asphalt cement be simulated in the laboratory?

8.7 Show how various Superpave tests used to characterize the asphalt binder are related to pavement performance.

8.8 Define the four methods used to grade asphalt binders. Which method is used in your state?

8.9 Discuss how asphalt emulsions work as a binder in asphalt mixtures.

8.10 What are the components of hot-mix asphalt? What is the function of each component in the mix?

8.11 What are the objectives of the asphalt concrete mix-design process?

8.12 Briefly describe the volumetric mix design procedure of Superpave.

References

The Asphalt Institute. *The Asphalt Handbook*. MS-4. Lexington, KY: The Asphalt Institute, 1989.

The Asphalt Institute. *Introduction to Asphalt*. MS-5. Lexington, KY: The Asphalt Institute, 1993.

The Asphalt Institute. *Mix Design Methods for Asphalt Concrete and Other Hot-Mix Types*. MS-2. Lexington, KY: The Asphalt Institute, 1995.

Goetz, W. H. and L. E. Wood. "Bituminous Materials and Mixtures". *Highway Engineering Handbook*, Section 18. New York: McGraw-Hill, 1960.

Huang, Y. H. *Pavement Analysis and Design*. 2d ed. Upper Saddle River, NJ: Prentice Hall, 2004.

Jansich, D. W., and F. S. Gaillard. *Minnesota Seal Coat Handbook*. Maplewood, MN: Department of Transportation, 1998.

Mamlouk, M. S. "Rheology of Cold-Recycled Pavement Materials Using Creep Test". *Jounal of Testing and Evaluation* 12(6):341-347, 1984.

Mamlouk, M. S. and J. P. Zaniewski. *Pavement Preventive Maintenance: Description, Effectiveness, and Treatments*. Special Technical Publication 1348. West Conshohocken, PA: American Society for Testing and Materials, 1998.

McGennis, R. B. et al. *Background of Superpave Asphalt Binder Test Methods*. Publication no. FHWA-SA-94-069. Washington, DC: Federal Highway Administration, 1994.

McGennis, R. B. et al. *Background of Superpave Asphalt Mixture Design and Analysis*. Publication no. FHWA-SA-95-003. Washington, DC: Federal Highway Administration, 1995.

Peterson, J. C. *Chemical Composition of Asphalt as Related to Asphalt Durability—State of the Art*. Transportation Research Record No. 999, Washington, DC: Transportation Research Board, 1984.

Roberts, F. L., P. S. Kandhal, E. R. Brown, D. Y. Lee, and T. W. Kennedy. *Hot Mix Asphalt Materials, Mixture Design, and Construction*. 2d ed. Lanham, MD: NAPA Education Foundation, 1996.

Witczak, M. W., K. Kaloush, T. Pellinen, M. El-Basyouny, and H. Von Quintus. *Simple Performance Test for Superpave Mix Design*. NCHRP Report 465. Washington, D. C: National Cooperative Highway Research Program, National Research Council, 2002.

CHAPTER 9
WOOD

Wood, because of its availability, relatively low cost, ease of use, and durability, if properly maintained, continues to be an important civil engineering material. Wood is used extensively for buildings, bridges, utility poles, floors, roofs, trusses, and piles. Civil engineering applications include both natural wood and engineered wood products, such as laminates, plywood, and strand board. In order to use wood efficiently, it is important to understand its basic properties and limitations. In the United States, the Forest Service of the Department of Agriculture has broad management responsibility for the harvesting of wood from public lands and for assisting private sources with the selection of products for harvesting. This agency has produced an excellent document describing the characteristics and properties of wood (USDA-FS, 1999).

This chapter covers the properties and characteristics of wood. In the design of a wood structure, joints and connections often limit the design elements. These are generally covered in a design class for wood construction and, therefore, are not considered in this text.

Wood is a natural, renewable product from trees. Biologically, a tree is a woody plant that attains a height of at least 6 m, normally has a single self-supporting trunk with no branches for about 1.5 m above the ground, and has a definite crown. There are over 600 species of trees in the United States.

Trees are classified as either endogenous or exogenous, based on the type of growth. Endogenous trees, such as bamboo, grow with intertwined fibers. Wood from endogenous trees is not generally used for engineering applications in the United States. Exogenous trees grow from the center out by adding concentric layers of wood around the central core. This book considers only exogenous trees.

Exogenous trees are broadly classified as deciduous and conifers, producing hardwoods and softwoods, respectively. The terms hardwood and softwood are classifications within the tree family, not a description of the woods' characteristics. In general, softwoods are softer, less dense, and easier to cut than hardwoods. However, exceptions exist such as Balsa, a very soft and lightweight wood that is technically a hardwood.

Deciduous trees generally shed their leaves at the end of each growing season. Commercial hardwood production in the U. S. comes from 40 different tree species. Hardwoods are generally used for furniture and decorative veneers, due to their pleasing grain pattern. The cost of hardwoods limits their construction application.

Conifers, also known as evergreens, have needlelike leaves and normally do not shed them at the end of the growing season. Conifers grow continuously through the crown, producing a uniform stem and homogenous characteristics(Panshin & De Zeeuw, 1980). Softwood production in the U. S. comes from about 20 individual species of conifers. Conifers are widely used for construction. Conifers grow in large stands, permitting economical harvesting. They mature rapidly, making them a renewable resource. Table 9.1 shows examples of hardwood and softwood species (USDA-FS, 1999).

Table 9.1 Major Sources of Hardwood and Softwood Species by Region

Hardwoods			Softwoods		
Western	Northern and Appalachia	Southern	Western	Northern	Southern
Ash	Ash	Red alder	Incense cedar	Northern white cedar	Atlantic white cedar
Basswood	Aspen	Oregon ash	Port Orford cedar	Balsam fir	Bald cypress
American beech	Basswood	Aspen	Douglas-fir	Eastern hemlock	Fraser fir
Butternut	Buckeye	Black cottonwood	Western hemlock	Fraser fir	Southern pine
Cottonwood	Butternut	California black oak	Western larch	Jack pine	Eastern red cedar
Elm	American beech	Big leaf maple	Lodgepole pine	Eastern white pine	
Hackberry	Birch	Paper birch	Ponderosa pine	Eastern red cedar	
Pecan hickory	Black cherry	Tan oak	Suger pine	Eastern spruces	
Ture hickory	American chestnut*		Western white pine	Tamarack	
Honey locust	Cottonwood		Western red cedar		
Black locust	Elm		Red wood		
Magnolia	Hackberry		Englemann spruce		
Soft maple	True hickory		Siitka spruce		
Red oak	Honey locust		Yellow cedar		
White oak	Black locust				
	Hard maple				

Hardwoods			Softwoods		
Western	Northern and Appalachia	Southern	Western	Northern	Southern
Sassafras	Soft maple				
Sweetgum	Red oak				
American sycamore	White oak				
	American sycamore				
Tupelo	Black walnut				
Black walnut	Yellow poplar				
Black willow					
Yellow poplar					

* Chestnut no longer harvested, but lumber from salvage timbers available.

9.1 Structure of Wood

Wood has a distinguished structure that affects its use as a construction material. Civil and construction engineers need to understand the way the tree grows and the anisotropic nature of wood in order to properly design and construct wood structures.

9.1.1 Growth Rings

The concentric layers in the stem of exogenous trees are called tree rings or annual rings, as shown in Figure 9.1a. The wood produced in one growing season constitutes a single growth ring. Each annual ring is composed of early wood, produced by rapid growth during the spring, and latewood from summer growth. Latewood consists of dense, dark, and thick-walled cells producing a stronger structure than early wood, as shown in Figure 9.1b. The predominant physical features of the tree stem include the bark, cambium, wood, and pith, as shown in Figure 9.2. The bark is the exterior covering of the tree and has an outer and an inner layer. The outer layer is dead and corky and has great variability in thickness, dependent on the species and age of the tree. The inner bark layer is the growth layer for bark but is not part of the wood section of the tree. The cambium is a thin layer of cells situated between the wood and the bark and is the location of all wood growth.

The wood section of the tree is composed of sapwood and heartwood. Sapwood functions as a storehouse for starches and as a pipeline to transport sap. Generally, faster growing spe-

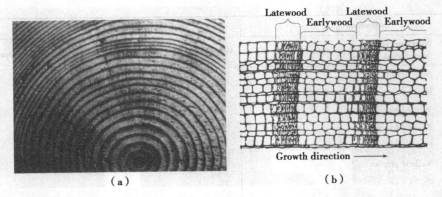

Figure 9.1 Cross section of a typical tree stem: (a) annual rings (photo courtesy of American Forest & Paper Association, Washington, D.C.) and (b) earlywood and latewood.

cies have thick sapwood regions. In its natural state, sapwood is not durable when exposed to conditions that promote decay. Heartwood is not a living part of the tree. It is composed of cells that have been physically and chemically altered by mineral deposits. The heartwood provides structural strength for the tree. Since the heartwood does not contain sap, it is naturally resistant to decay.

The pith is the central core of the tree. Its size varies with the tree species, ranging from barely distinguishable to large and conspicuous. The color ranges from blacks to whitish, depending on the tree species and locality. The pith structure can be solid, porous, chambered, or hollow.

9.1.2 Anisotropic Nature of Wood

Wood is an anisotropic material in that it has different and unique properties in each direction. The three axis orientations in wood are longitudinal or parallel to the grain, radial or cross the growth rings, and tangential or tangent to the growth rings, as illustrated in Figure 9.2. The anisotropic nature of wood affects physical and mechanical properties such as shrinkage, stiffness, and strength.

The anisotropic behavior of wood is the result of the tubular geometry of the wood cells. The wood cells have a rectangular cross section. The centers of the tubes are hollow, whereas the ends of the tubes are tapered. The length-to-width ratio can be as large as 100. The long dimension of the majority of cells is parallel to the tree's trunk. However, a few cells, in localized bundles, grow radially, from the center to the outside of the trunk. The preponderance of cell orientation in one direction gives wood its anisotropic characteristics. The hollow tube struc-

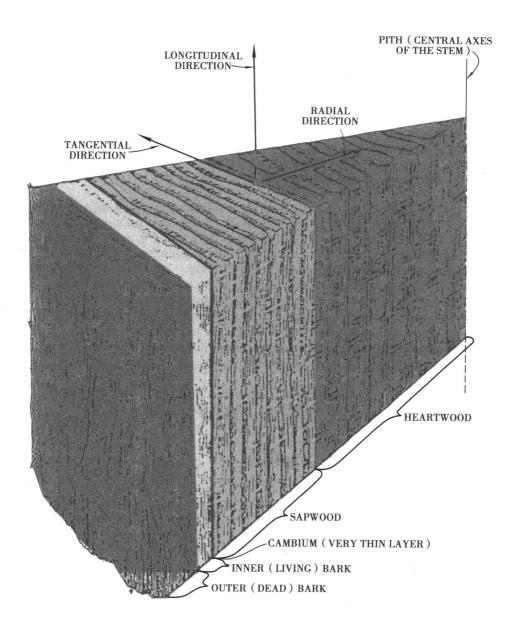

Figure 9.2 Main structural features of a typical tree stem.

ture is very efficient in resisting compressive stresses parallel to its length, but readily deforms when loaded on its side. Also, fluctuations in moisture contents flex the tube walls but have little effect on the length of the tube.

9.2 Chemical Composition

Wood is composed of cellulose, lignin, hemicellulose, extractives, and ash producing minerals. Cellulose accounts for approximately 50 percent of the wood substance by weight(USDA-FS, 1999). The exact percent is species dependent. It is a linear polymer(aliphatic carbon compound)having a high molecular weight. The main building block of cellulose is sugar-glucose. As the tree grows, linear cellulose molecules arrange themselves into highly ordered strands, called *fibrils*. These ordered strands form the large structural elements that compose the cell walls of wood fibers.

Lignin accounts for 23% to 33% of softwood and 16% ~ 25% of hard-wood by weight. Lignin is mostly an intercellular material. Chemically, lignin is an intractable, insoluble, material that is loosely bonded to the cellulose. Lignin is basically the glue that holds the tubular cells together. The longitudinal shear strength of wood is limited by the strength of the lignin bounds.

Hemicelluloses are polymeric units made from sugar molecules. Hemi-cellulose is different from cellulose in that it has several sugars tied up in its cellular structure. Hardwood contains 20% to 30% hemicellulose and softwood averages 15% to 20%. The main sugar units in hardwood and softwood are xylose and monnose, respectively.

The extractives compose 5% to 30% of the wood substance. Included in this group are tannins and other polyphenolics, coloring matters, essential oils, fats, resins, waxes, gums, starches, and simple metabolic intermediates. These materials can be removed with simple inert neutral solvents, such as water, alcohol, acetone, and benzene. The amount contained in an individual tree depends on the species, growth conditions, and time of year the tree is harvested.

The ash-forming materials account for 0.1% ~ 3.0% of the wood material and include calcium, potassium, phosphate, and silica.

9.3 Physical Properties

Important physical properties include specific gravity and density, thermal properties, and electrical properties.

9.3.1 Specific Gravity and Density

Specific gravity of wood depends on cell size, cell wall thickness, and number and types of cells. Regardless of species, the substance composing the cell walls has a specific gravity of 1.5. Because of this consistency, specific gravity is an excellent index for the amount of substance a dry piece of wood actually contains and is nearly constant within each species. Therefore, specific gravity, or density, is a commonly cited property and is an indicator of mechanical properties within a clear, straight-grained wood.

The dry density of wood ranges from 160 kg/m^3 (10 pcf) for balsa to 1000 kg/m^3 (65 pcf) for some species. The majority of wood types have densities in the range of 300 to 700 kg/m^3 (20 to 45 pcf). Within common domestic species, density may vary by $\pm 10\%$.

9.3.2 Thermal Properties

Thermal conductivity, specific heat, thermal diffusivity, and coefficient of thermal expansion are the four significant thermal properties of wood.

Thermal Conductivity Thermal conductivity is a measure of the rate at which heat flows through a material. The reciprocal of thermal conductivity is the thermal resistance (insulating) value (R). Wood has a thermal conductivity that is a fraction of that of most metals and three to four times greater than common insulating materials. The thermal conductivity ranges from 0.06 W/(mK) [0.34 Btu/(h-ft-°F)] for balsa to 0.17 W/(mK) [1.16 Btu/(h-ft-°F)] for rock elm. Structural woods average 0.12 W/(mK) [0.07 Btu/(h-ft-°F)] as compared to 200 W/(mK) [115 Btu/(h-ft-°F)] for aluminum and 0.04 W/(mK) [0.025 Btu/(h-ft-°F)] for wool. The thermal conductivity of wood depends on several iterns including (1) grain orientation, (2) moisture content, (3) specific gravity, (4) extractive content, and (5) structural irregularities such as knots.

Heat flow in wood across the radial and tangential directions (with respect to the growth rings) is nearly uniform. However, heat flow through wood in the longitudinal direction (parallel to the grain) is 2.0 to 2.8 times greater than in the radial direction.

Moisture content has a strong influence on thermal conductivity. When the wood is dry, the cells are filled with air and the thermal conductivity is very low. As the moisture content increases, thermal conductivity increases. As the moisture content increases from 0% to 40%, the thermal conductivity increases by about 30%.

Because of the solid cell wall material in heavy woods, they conduct heat faster than light woods. This relationship between specific gravity and thermal conductivity for wood is linear. Also affecting the heat transfer in wood are increases in extractive content and density(i.e., knots) which increase thermal conductivity.

Specific Heat Specific heat of a material is the ratio of the quantity of heat required to raise the temperature of the material one degree to that required to raise the temperature of an equal mass of water one degree. Temperature and moisture content largely control the specific heat of wood, with species and density having little to no effect. When wood contains water, the specific heat is increased because the specific heat of water is higher than that of dry wood. However, the value of specific heat for the wet wood is higher than just the sum of the specific heats for the wood and water combined. This increase in specific heat beyond the simple sum is due to the wood-water bonds absorbing energy. An increase in temperature increases the energy absorption of wood and results in an increase in the specific heat.

Thermal Diffusivity Thermal diffusivity is a measure of the rate at which a material absorbs heat from its surroundings. The thermal diffusivity for wood is much smaller than that of other common building materials. Generally, wood has a thermal diffusivity value averaging 0.006 mm/sec., compared with steel, which has a thermal diffusivity of 0.5 mm/sec. It is because of the low thermal diffusivity that wood does not feel hot or cold to the touch, compared with other materials. The small thermal conductivity, moderate density, and moderate specific heat contribute to the low value of thermal diffusivity in wood.

Coefficient of Thermal Expansion The coefficient of thermal expansion is a measure of dimensional changes caused by a temperature variance. Thermal expansion coefficients for completely dry wood are positive in all directions. For both hard and soft woods, the longitudinal(parallel to the grain) coefficient values range from 0.009 to 0.0014 mm/m/℃. The expansion coefficients are proportional to density and therefore are five to ten times greater across the grain than those parallel to it.

When moist wood is heated, it expands due to thermal expansion and then shrinks because of the loss of moisture (below the fiber saturation point). This combined swelling and shrinking often results in a net shrinkage. Most woods, at normal moisture levels, react in this way.

9.3.3 Electrical Properties

Air-dry wood is a good electrical insulator. As the moisture content of the wood increases, the resistivity decreases by a factor of three for each 1% change in moisture content. However, when wood reaches the fiber saturation point, it takes on the resistivity of water alone.

9.4 Mechanical Properties

Knowing the mechanical properties of wood is a prerequisite to a proper design of a wood structure. Typical mechanical properties of interest to civil and construction engineers include modulus of elasticity, strength properties, creep, and damping capacity.

9.4.1 Modulus of Elasticity

The typical stress-strain relation of wood is linear up to a certain limit, followed by a small nonlinear curve after which failure occurs, as shown on Figure 9.3. The modulus of elasticity of wood is the slope of the linear portion of the representative stress-strain curve. The stress-strain relation of wood varies within and between species and is affected by variation in moisture content and specific gravity. Also, since wood is anisotropic, different stress-strain relations exist for different directions. The moduli of elasticity along the longitudinal, radial, and tangential axes are typically diferent.

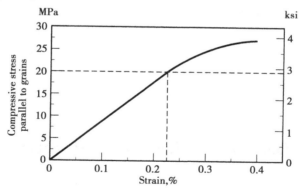

Figure 9.3 Typical stress strain relationship for wood.

9.4.2 Strength Properties

Strength properties of wood vary to a large extent, depending on the orientation of grain relative to the direction of force. For example, the tensile strength in the longitudinal direction (parallel to grain) is more than 20 times the tensile strength in the radial direction (perpendicular to grain). Also, tensile strength in the longitudinal direction is larger than the compressive strength in the same direction. Common strength properties for wood include modulus of rupture in bending, compressive strength parallel and perpendicular to the grain, and shear strength parallel to the grain. Some of the less common strength properties are tensile strength parallel to the grain, torsion, toughness, fatigue strength, and rolling shear strength.

9.4.3 Creep

Under sustained loads wood continues to deform or creep. The design values of material properties contemplate fully stressing the member to the tabulated design values for a period of 10 years and/or the application of 90% of the full maximum load continuously throughout the life of the structure. If the maximum stress levels are exceeded, the structure can deform prematurely.

9.4.4 Damping Capacity

Damping is the phenomenon in which the amplitude of vibration in a material decreases with time. Reduction in amplitude is due to internal friction within the material and resistance of the support system. Moisture content and temperature largely govern the internal friction in wood. At normal ambient temperatures, an increase in moisture content produces a proportional increase in internal friction up to the fiber-saturation point. Under normal conditions of temperature and moisture content, the internal friction in wood (parallel to the grain) is 10 times that of structural metals. Because of these qualities, wood structures dampen vibrations more quickly than metal structures of similar design.

9.5 Testing to Determine Mechanical Properties

Standard mechanical testing methods for wood are designed almost exclusively to obtain data for predicting performance. To achieve reproducibility in the testing environment, specifications include methods of material selection and preparation, testing equipment and techniques, and computational methods for data reduction. Standards for testing wood and wood composites are published by ASTM, the U. S. Department of Commerce, the National Standard Institute(NSI), and various other trade associations, such as the Western Woods Product Association.

Due to the many variables affecting the test results, it is of primary importance to correctly select the specimen and type of test. There are two main testing techniques for establishing strength parameters: the testing of representative, small, clear specimens and the testing of timbers of structural sizes.

The primary purposes for testing small, clear specimens are to obtain the machanical properties of various species and to provide a means of control and comparison in production activities. The testing of structural timbers provides relationships among mechanical and physical properties, working stress data, correlations between environmental conditions, wood imperfections, and mechanical properties. ASTM D143 presents the complete testing standards for small, clear wood specimens. This standard gives full descriptions of sample collection, preparation, and testing techniques. Mechanical tests included in this standard are the following:

- Static bending
- Impact bending
- Compression perpendicular to the grain
- Shear parallel to the grain
- Tension parallel to the grain
- Nail withdrawal
- Radial and tangential shrinkage
- Compression parallel to the grain
- Toughness
- Hardness
- Cleavage
- Tension perpendicular to the grain
- Specific gravity and shrinkage in volume
- Moisture determination

Figure 9.4 shows a schematic of test specimens of wood tested in tension, compression, bending, and hardness. Static and impact bending, compression and tension parallel and perpendicular to the grain, and shear parallel to the grain are commonly used.

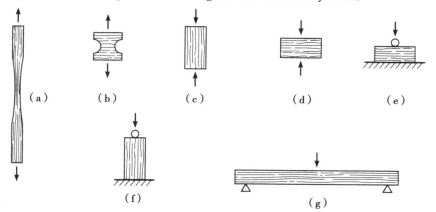

Figure 9.4 Test specimens of wood: (a) tension parallel to grains, (b) tension perpendicular to grains, (C) compression parallel to grains, (d) compression perpendicular to grains, (e) hardness perpendicular to grains, (f) hardness parallel to grains, and (g) bending. (©Pearson Education, Inc. Used by permission.)

9.5.1 Static Bending Test

The static bending test is performed on either 50 mm×50 mm×760 mm or 25 mm×25 mm× 410 mm specimens. For the large specimens, the loading head is placed on the center of the specimen and over a span of 710 mm, and the load is applied at a rate of 2.5 mm/min. For the small specimens, the loading head is placed on the center of the specimen and over a span of 360 mm, and the load is applied at a rate of 1.3 mm/min. Load-deflection data are recorded to or beyond the maximum load. Within the proportional limit, readings are taken to the nearest 0.02 mm. After the proportional limit, deflection readings are usually measured with a dial gage, to the limit of the gage, usually 25 mm. Load and deflection of the first failure, the maximum load, and points of sudden change are recorded. The failure appearance is described as either brash or fibrous. Brash indicates an abrupt failure and fibrous indicates a failure showing splinters.

Sample Problem 9.1

A static bending test was performed on a 50 mm×50 mm×760 mm wood sample according to ASTM D143 procedure (span between supports=710 mm). If the maximum load was 2.67 kN, calculate the modulus of rupture.

Solution

$$\text{Modulus of rupture} = \frac{Mc}{I}$$

where

M = bending moment at maximum load

c = 1/2 of the specimen height

I = moment of inertia of the specimen cross section

Reaction at each support at failure = $\frac{2.67}{2}$ = 1.335 kN

Bending moment at the center at failure = $1.335 \times \left(\frac{710}{2}\right)$ = 473.9 N·m

$$I = \frac{(0.05)(0.05)^3}{12} = 5.21 \times 10^{-7} \, m^4$$

Modulus of rupture = $\frac{473.9 \times 0.025}{5.21 \times 10^{-7}}$ = 22.75 MPa

9.5.2 Compression Tests

The compression test parallel to the grain is performed on either 50 mm×50 mm×200 mm or 25 mm×25 mm×100 mm specimens, as shown in Figure 9.5. The load is applied at a rate equal to 0.003 mm/mm of the nominal specimen length per minute. The deformations are recorded to 0.002 mm over a gage length of not more than 150 mm for the large specimens or 50 mm for the small specimens. Load-compression readings are recorded until well past the proportional limit. The failures should occur in the center portion of the sample. If failures are occurring near the ends, the samples can be stacked such that the ends dry relative to the middle. This will increase the strength of the ends of the sample. The tests are then repeated on the conditional samples. The type of failure can be classified as crushing, wedge split, shearing, splitting, compression and shearing, and brooming and end rolling, as shown in Figure 9.6.

Figure 9.5 Compression parallel to grain test.

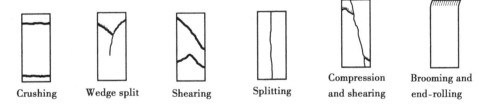

Crushing   Wedge split   Shearing   Splitting   Compression and shearing   Brooming and end-rolling

Figure 9.6 Types of failure in the compression parallel-to-grain test. (ASTM D143). Reprinted with permission of ASTM.

The compression test perpendicular to the grain is performed on 50 mm×50 mm×150 mm specimens. The load is applied through a metal bearing plate 50 mm in width, centered across the upper surface of the specimen. The load is applied at a rate of 0.305 mm/min. Deflection readings are taken to the nearest 0.002 mm. Load and deformation are measured until the deformation is 2.5 mm.

SUMMARY

Wood is an extremely flexible building material. Historically, natural wood products were the only option available to the engineer. However, modern forestry practices limit the size of natural products that are available. The need to increase the efficiency of using wood products has led to the development of engineered wood products. These products are frequently more economical than natural wood, particularly when large dimensions are required. In addition, by careful control of the manufacturing process, engineered woods can be produced that have characteristics superior to natural wood. The characteristics of engineered wood products depend on the wood stock used, the quality of the adhesive, and the manufacturing process. Several factors make wood unique when compared to the other materials used in civil engineering, including anisotropy, moisture sensitivity, creep, and the existence of defects in wood products. Furthermore, when wood is exposed to the environment, care must be taken to prevent degradation due to fungi, bacteria, and insects.

QUESTIONS AND PROBLEMS

9.1 What is the difference between early wood and latewood? Describe each.

9.2 A simple lab test for specific gravity on two samples of lumber indicate that sample A has $G = 1.12$ and sample B has $G = 1.03$. Based on this information alone, which wood sample would you choose as a structural member for your construction project. Briefly explain why.

9.3 Discuss the anisotropic nature of wood. How does this phenomenon affect the performance of wood?

9.4 Briefly describe the chemical composition of wood.

9.5 What are the factors considered in grading lumber? What are the main grades of hardwoods and softwoods?

9.6 State five different imperfections that may be found in lumber, and briefly define them.

9.7 List five different tests used to evaluate the mechanical properties of wood.

9.8 A short rounded wood column with a diameter of 10 inches is to be constructed. If the failure stress is 7.3 ksi, what is the maximum load that can be applied to this column, using a factor of safety of 1.3?

9.9 For the purpose of designing wood structures, laboratory, measured strength properties are adjusted for application conditions. State five different application conditions that are used to adjust the strength properties.

9.10 What are the five types of organisms that attack wood?

9.11 What are the two types of preservatives that can be used to protect wood from decay? How are these preservatives applied?

9.12 What are the main types of engineered wood products?

9.13 What are the main advantages of engineered wood products over natural-timber members?

References

American Forest & Paper Association, American Wood Council. *National Design Specification for Wood Construction*. Washington. D. C; American Forest & Paper Association, American Wood Council, 2005.

APA-EWS, PRL-501. *Performance Standard for APA EWS Laminated Veneer Lumber*. Tacoma, WA: APA-The Engineered Wood Association, 2000.

APA-EWS, *Glulam Design Properties and Layup Combinations*. Tacoma, WA: APA-The Engineered Wood Association, 2004.

 e-builder. http://www. ebuild. com/guide/resources/product-news. asp? ID=81435, 2005.

Forintek. *Borate-Treated Wood for Construction*. Forintek Canada Corp.

 http://durable-wood. com/pdfs/borate-engloct 02. pdf, 2002.

Levin, E. , ed. *The International Guide to Wood Selection*. New York. NY: Drake Publishers, Inc, 1972.

Panshin, A. J. and C. De Zeeuw. *Textbook of Wood Technology*. 4th ed. New York. NY: McGraw-Hill Book Company, 1980.

Southern Pine Council, Southern Pine Products. Association. *Pressure Treated Southern Pine*. Kenner, LA: Southern Pine Council, Southern Pine Products Association, 1990.

Trusjoist. *Specifier's Guide, Parallam PSL Headers, Bearns, and Columns*. Boise, ID: Trus Joist, a Weyerhaeuser Business.

 http://www. trusjoist. com/PDFFiles/2060. pdf, 2003.

USDA-FS. *Wood Handbook—Wood as an Engineering Material*. Gen. Tech. Rep. FPL-GTR-113. Madison, WI: USDA, Forest Service, Forest Products Laboratory, 1999. http://www. fpl. fs. fed. us/documnts/fplgtr/fplgtr113/fplgtr113. htm, 1999.

Van Vlack, L. H. *Materials for Engineering : Concepts and Applications*, Reading, MA: Addison-Wesley, 1982.

Van Vlack, L. H. *Elements of Materials Science and Engineering*, 6th ed. Reading. MA: Addison Wesley, 1989.

Williamson, T. G. (ed.). *APA Engineered Wood Handbook*. New York, NY: McGraw-Hill Book Company, 2002.

APPENDIX

LABORATORY MANUAL

The laboratory tests discussed in this appendix can be performed as a part of the civil and construction engineering materials course. More tests are included in this manual than typically can be performed in one semester. The extent of tests provided gives the instructor flexibility to choose the appropriate tests. In order for the students to get the most benefit from the laboratory sessions, tests should be coordinated with the topics covered in the lectures.

This laboratory manual summarizes the main components of each test method. Students are encouraged to read the corresponding ASTM or AASHTO test methods for the detailed laboratory procedures. The ASTM and AASHTO standards are usually available in college libraries.

In many cases, the time available for a laboratory session is not sufficient to permit the performance of the complete test, as specified in the ASTM or AASHTO procedure. Therefore, the instructor may limit the number of specimens or eliminate some portions of the test. However, the student should be aware of the complete test procedure and the specification requirements.

In some cases, different experiments are used to obtain the same material properties, such as the air content of freshly mixed concrete. In such cases, the instructor may require the specific test used by the state or the test for which equipment is available.

Typically a laboratory report is required by the student after each laboratory session. The format of the report can vary, depending on the requirements of the instructor. A suggested format may include the following items:

- Title of the experiment
- ASTM or AASHTO designation
- Purpose
- Significance and use
- Test materials
- Main pieces of equipment
- Summary of test procedure and test conditions
- Test results and analysis
- Comments, conclusions, and recommendations. Any deviations from the standard test procedure should be reported and justified.

1. Introduction to Measuring Devices

ASTM Designation

There is no ASTM procedure for the main portion of this session. The information at the end of Chapter 1 can be used as a reference. Some discussion on precision and bias can be helpful; this is included in ASTM C670 (Practice for Preparing Precision and Bias Statements for Test Methods for Construction Materials).

Purpose

To introduce the students to common measuring devices, such as dial gauges, LVDTs, strain gauges, proving rings, extensometers, etc. An introduction to precision and bias can also be included.

Apparatus

The instructor may demonstrate one or more of the following items:
- A few dial gauges with different ranges and sensitivities (Figure 1.23)
- LVDT (Figures 1.25-1.27) and necessary attachments such as power supply, signal conditioner, voltmeter, display device, and calibration device
- Extensometer (Figures A.1, 1.24 and 1.28)
- Strain gauge (Figure 1.29) and necessary attachments

266 Materials for Civil and Construction Engineers

◆ Proving ring(Figures A. 2 and 1. 30)
◆ Load cell(Figures A. 3 and 1. 31)

Figure A. 1   Extensometers.

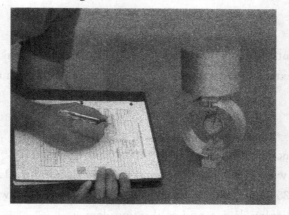

Figure A. 2   Calibrating a proving ring using a known-weight object.

Appendix 267

Figure A.3 Load cells.

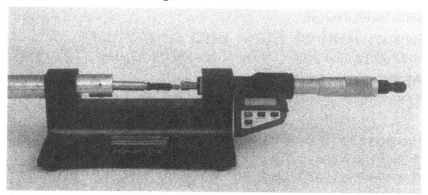

Figure A.4 Calibrating an LVDT with a micrometer using a calibration device.

Calibration

The instructor may require the students to calibrate one or more measuring devices, such as a proving ring or an LVDT. Static loads can be used to calibrate the proving ring to develop a relation between force and the reading of the proving ring (Figure A. 2). To calibrate the LVDT, a calibration device equipped with a micrometer, such as that shown in Figures A. 4 and 1. 26 is used. A relation is developed between voltage and displacement of the LVDT.

Requirements

- Brief description of the demonstrated device(s) including the use, components, theory, sensitivity, etc.
- Calibration table, graph, and equation for each device calibrated

2. Tension Test of Steel and Aluminum

ASTM Designation

ASTM E8—Tension Testing of Metallic Materials

Purpose

- To determine stress-strain relationship
- To determine yield strength
- To determine tensile strength
- To determine elongation and reduction of cross-sectional area
- To determine modulus of elasticity
- To determine rupture strength

Significance and Use

This test provides information on strength and ductility for metals subjected to a uniaxial tensile stress. This information may be useful in comparison of materials, alloy developments, quality control, design under certain circumstances, and detecting nonuniformity and imperfections, as indicated by the fracture surface.

Apparatus

- A testing machine capable of applying tensile load at a controlled rate of deformation or load. The testing machine could be either mechanical or closed-loop electrohydraulic. The machine could be equipped with a dial gauge to indicate the load or could be connected to a chart recorder or computer to record load and deformation.

- ◆ A gripping device, used to transmit the load from the testing machine to the test specimen and to ensure axial stress within the gauge length of the specimen
- ◆ An extensometer with an LVDT or dial gauge used to measure the deformation of the specimen
- ◆ Caliper to measure the dimensions of the specimen

Test Specimens

Either plate-type or rounded specimens can be used, as shown in Figures A.5 and 2.9. Specimen dimensions are specified in ASTM E8.

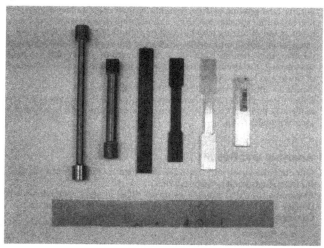

Figure A.5 Rounded and plate-type steel and aluminum tension test specimens.

Test Procedure

1. Mark the gauge length on the specimen, either by slight notches or with ink.
2. Place the specimen in the loading machine (Figure 2.10).
3. Attach the extensometer to the specimen (Figure A.6).
4. Set the load reading to zero, then apply load at a rate less than 690 kPa/min. Unless otherwise specified, any convenient speed of testing may be used up to half of the specified yield strength or yield point, or one quarter of the specified tensile strength, whichever is smaller. After the yield strength or yield point has been determined, the strain rate may be in-

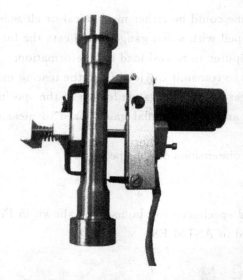

Figure A.6 Extensometer attached to a steel specimen.

creased to a maximum of 0.5 in./in. of the gauge length per minute.
5. Continue applying the load until the specimen breaks.
6. Record load and deformation every 2.2 kN (500 lb) increment for steel and every 890 N increment for aluminum, both before and after the yield point.

Analysis and Results

- Calculate the stress and strain for each load increment until failure.
- Plot the stress versus strain curve.
- Determine the yield strength using the offset method, extension method (Figure 1.7), or by observing the sudden increase in deformation.
- Calculate the tensile strength

$$\sigma = P_{max}/A_o$$

where

σ = tensile strength, MPa (psi)
P_{max} = maximum load carried by the specimen during the tension test, N (lb)
A_o = original cross-sectional area of the specimen, mm² (in.²)

- Calculate the elongation

Percent elongation = $[(L_s - L_o)/L_o] \times 100$

where

L_s = gage length after rupture, mm (in.)

L_o = original gage length, mm(in.)

For elongation > 3.0%, fit the ends of the fractured specimen together and measure L_s as the distance between two gauge marks. For elongation ≤ 3.0%, fit the fractured ends together and apply an end load along the axis of the specimen sufficient to close the fractured ends together, then measure L_s as the distance between gauge marks.

◆ Calculate the modulus of elasticity

$$E = \sigma/\varepsilon$$

where

E = modulus of elasticity, MPa(psi)
σ = stress in the proportional limit, MPa(psi)
ε = corresponding strain, mm/mm(in./in.)

◆ Calculate the rupture strength

$$\sigma_r = P_f/A_o$$

where

σ_r = rupture strength, MPa(psi)
P_f = final load, N(lb)
A_o = original cross-sectional area, mm² (in.²)

◆ Calculate the reduction of cross-sectional area

Percent reduction in cross-sectional area = $[(A_o - A_s)/A_o] \times 100$

where

A_s = cross section after rupture, mm² (in.²)

To calculate the cross section after rupture, fit the ends of the fractured specimen together and measure the mean diameter or width and thickness at the smallest cross section.

Replacement of Specimens

The test specimen should be replaced if
◆ the original specimen had a poorly machined surface.
◆ the original specimen had wrong dimensions.
◆ the specimen's properties were changed because of poor machining practice.
◆ the test procedure was incorrect.
◆ the fracture was outside the gauge length.
◆ for elongation determination, the fracture was outside the middle half of the gauge length.

Report

◆ Stress-strain relationship

- Yield strength and the method used
- Tensile strength
- Elongation and original gauge length
- Modulus of elasticity
- Rupture strength
- Reduction of cross-sectional area

3. Impact Test of Steel

ASTM Designation
ASTM E23—Test Methods for Notched Bar Impact Testing of Metallic Materials

Purpose
To determine the energy absorbed in breaking notched steel specimens at different temperatures, using the Charpy V Notch test. The energy value is a measure of toughness of the material.

Significance and Use
This test measures a specimen's change in toughness as the temperature changes.

Apparatus
Use an impact testing machine of the pendulum type of rigid construction and of capacity more than sufficient to break the specimen in one blow (Figures A.7 and 2.15).

Test Specimen
Steel specimens prepared according to ASTM E23, as shown in Figure 2.9.

Test Conditions
The test can be performed at the following four temperatures:
- -40 ℃ (-40 °F)(dry ice+isopropyl alcohol)(dry alcohol)
- -18 ℃ (0 °F)(dry ice+30% isopropyl alcohol+70%water)
- 4 ℃ (40 °F)(ice+water)
- 40 ℃ (104 °F)(oven)

Figure A.7 Charpy V notch impact testing machine.

Test Procedure

1. Prepare the impact testing machine by lifting up the pendulum and adjusting the gauge reading to zero. Since the pendulum is heavy, be careful to handle it safely.
2. Remove the test specimen from the temperature medium using tongs, and immediately position it on the two anvils in the impact testing machine.
3. Release tlle pendulum without vibration by pushing the specified button. The time between removing the specimen from the temperature medium and the completion of the test should be less than 5 seconds.
4. Record the energy required to break the specimen by reading the gauge mark.
5. Observe the fracture surface appearance (Figure 2.11).
6. Measure the lateral expansion of the specimen using a caliper or the lateral expansion gauge specified in ASTM E23.

Report

- Energy required to cause fracture versus temperature plot
- Ductile-to-brittle transition temperature
- Fracture surface appearance (each specimen and temperature)
- Lateral expansion (each specimen and temperature)

4. Sieve Analysis of Aggregates

ASTM Designation
ASTM C136—Sieve Analysis of Fine and Coarse Aggregates

Purpose
To determine the particle size distribution of fine and coarse aggregate by dry sieving.

Significance and Use
This test is used to determine the grading of materials that are to be used as aggregates. It ensures that particle size distribution complies with applicable requirements and provides the data necessary to control the material of various aggregate products and mixtures containing aggregates. The data may also be useful in developing relationships concerning porosity and packing.

Apparatus
- Balances or scales with a minimum accuracy of 0.5 g for coarse aggregate or 0.1 g for fine aggregate
- Sieves
- Mechanical sieve shaker(Figures A.8, A.10, and 4.8)
- Oven capable of maintaining a uniform temperature of 110 ± 5 ℃ (230 ± 9 ℉)
- Sample splitter to reduce the quantity of the material to the size required for sieve analysis

Test Specimens
Thoroughly mix the aggregate sample and reduce it to an amount suitable for testing, using a sample splitter or by quartering. The minimum sample size should be as follows:

	Minimum Mass, kg
Fine aggregate with at least 95% passing 2.36-mm(No. 8) sieve	0.1
Fine aggregate with at least 85% passing 4.75-mm(No. 4) sieve	0.5
Coarse aggregate with a nominal maximum size of 9.5 mm(No. 3/8 in.)	1
Coarse aggregate of a nominal maximum size of 12.5 mm(1/2 in.)	2
Coarse aggregate of a nominal maximum size of 19.0 mm(3/4 in.)	5
Coarse aggregate of a nominal maximum size of 25.0 mm(1 in.)	10
Coarse aggregate of a nominal maximum size of 37.5 mm (1-1/2 in.)	15

Figure A.8 Table-top sieve shaker and sieves.

Figure A.9 Hanging-type sieve shaker and sieves for small samples of aggregates.

Test Procedure

1. Dry the aggregate test sample to a constant weight at a temperature of 110 ± 5 ℃, than cool to room temperature.
2. Select suitable sieve sizes to furnish the information required by the specifications covering the material to be tested. Common sieves in millimeters are 37.5, 25, 19, 12.5, 9.5, 4.75, 2.36, 1.18, 0.6, 0.3, 0.15, and 0.075 mm ($1\frac{1}{2}$ in., 1 in., $\frac{3}{4}$ in., $\frac{1}{2}$ in., $\frac{3}{8}$ in., NO. 4, NO. 8, No. 16, No. 30, No. 50, No. 100, and No. 200).
3. Nest the sieves in order of decreasing size of opening, and place the aggregate sample on the top sieve.
4. Agitate the sieves by hand or by mechanical apparatus for a sufficient period. The criterion for sieving time is that, after completion, not more than 1% of the residue on any individual sieve will pass that sieve during 1 minute of continuous hand sieving.
5. Determine the weight of each size increment.
6. The total weight of the material after sieving should be compared with the original weight of the sample placed on the sieves. If the amounts differ by more than 0.3%, based on the original dry sample weight, the results should not be used for acceptance purposes.

Analysis and Results

1. Calculate percentages passing, total percentages retained, or percentages of various sizes of fractions to the nearest 0.1%, on the basis of the total weight of the initial dry sample.
2. Plot the grain size distribution on a semilog graph paper (Figure A.10).
3. Plot the grain size distribution on a 0.45 power graph paper (Figure A.11).
4. Calculate the fineness modulus.

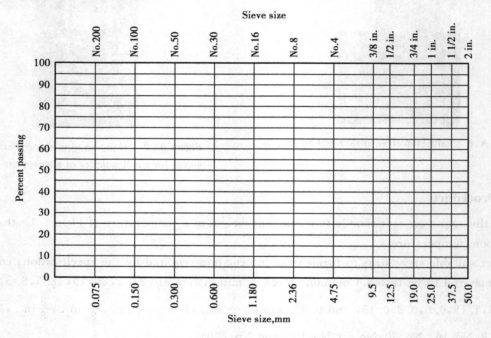

Figure A.10 Semi-log aggregate gradation chart.

Report

◆ Percentage of material retained between consecultive sieves, cumulative percentage of material retained on each sieve, or percentage of material passing each sieve. Report percentages to the nearest whole number, except if percentage passing 0.075 mm (No. 200) sieve is less than 10%, it should be reported to the nearest 0.1%.
◆ Grain size distribution plots using both semilog and 0.45 power gradation charts.
◆ Fineness modulus to the nearest 0.01.

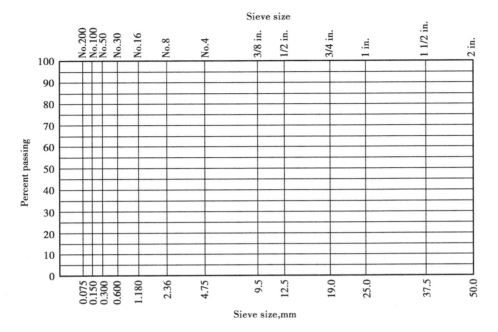

Figure A.11 0.45 power gradation chart.

5. Specific Gravity and Absorption of Coarse Aggregate

ASTM Designation

ASTM C127—Specific Gravity and Absorption of Coarse Aggregate

Purpose

To determine the specific gravity and absorption of coarse aggregate. The specific gravity may be expressed as bulk specific gravity, bulk specific gravity SSD(saturated-surface dry), or apparent specific gravity.

Significance and Use

Bulk specific gravity is generally used for the calculation of the volume occupied by the aggregate in various mixtures containing aggregates, including portland cement concrete, bituminous concrete, and other mixtures that are proportioned or analyzed on an absolute volume basis. Bulk specific gravity SSD is used if the aggregate is wet. Absorption values are used to calculate the change in the weight of aggregate due to water absorbed in the pore spaces within the constituent particles, compared with the dry condition.

Apparatus

- Balance accurate to 0.05% of the sample weight or 0.5 g, whichever is greater
- Wire basket 3.35 mm(No. 6) or finer mesh
- Water tank
- 4.75-mm(No. 4) sieve or other sizes as needed

Test Specimens

- Thoroughly mix the aggregate sample and reduce it to the approximate quantity needed, using an aggregate sample splitter or by quartering.
- Reject all materials passing 4.74 mm sieve by dry sieving and thoroughly washing to remove dust or other coatings from the surface.
- The minimum weight of test specimen to be used depends on the nominal maximum size as follows:

Nominal Maximum Size, mm	Minimum Mass, kg
12.5	2
19.0	3
25.0	4
37.5	5

Test Procedure

1. Immerse the aggregate in water at room temperature for a period of 24±4 hours.
2. Remove the test specimen from water and roll it in a large absorbent cloth until all visible films of water are removed. Wipe the larger particles individually.
3. Weigh the test sample in saturated surface-dry condition, and record it as B. Record this weight and all subsequent weights to the nearest 0.5 g or 0.05 % of the sample weight, whichever is greater.
4. Place the specimen in the wire basket and determine its weight while it is submerged in water at a temperature of 23±1.7 ℃, and record it as C. Take care to remove all entrapped air before weighing it by shaking the container while it is immersed.

5. Dry the test sample to a constant weight at a temperature of 110 ± 5 ℃, and weigh it and record this weight as A.

Analysis and Results

1. Bulk specific gravity $=A/(B-C)$
    where
    $A=$ mass of oven-dry sample in air, g.
    $B=$ mass of saturated surface-dry sample in air, g.
    $C=$ mass of saturated sample in water, g.
2. Bulk specific gravity (SSD) $=B/(B-C)$
3. Apparent specific gravity $=A/(A-C)$
4. Absorption, $\%=[(B-A)/A]\times100$

Report

- Bulk specific gravity
- Bulk specific gravity SSD
- Apparent specific gravity
- Absorption

6. Specific Gravity and Absorption of Fine Aggregate

ASTM Designation

ASTM C128—Specific Gravity and Absorption of Fine Aggregate

Purpose

To determine the specific gravity and absorption of fine aggregate. The specific gravity may be expressed as bulk specific gravity, bulk specific gravity SSD (saturated-surface dry), or apparent specific gravity.

Significance and Use

Bulk specific gravity is the characteristic generally used for calculating the volume occupied by the aggregate in various mixtures, including portland cement concrete, bituminous concrete, and other mixtures that are proportioned or analyzed on an absolute volume basis.

Apparatus

- Balance or scale with a capacity of 1 kg or more, sensitive to 0.1 g or less, and accurate within 0.1% of the test load
- Pycnometer or other suitable container into which the fine aggregate test sample can be readily introduced. A volumetric flask of 500 cm^3 capacity with a pycnometer top is satisfactory for a 500-g test sample of most fine aggregates (Figure A.12).
- Mold in the form of a frustum of a cone
- Tamper having a mass of 340 ± 15 g

Figure A.12 Mold, tamper, and volumetric flask used to determine bulk specific gravity and absorption of fine aggregate.

Test Procedure

1. Measure the weight of the pycnometer filled with water to the calibration mark. Record the weight as B.
2. Obtain approximately 1 kg of the fine aggregate sample.
3. Dry the aggregate sample in a suitable pan to constant weight at temperature of 110 ± 5 °C (230 ± 9 °F) and allow it to cool; then cover it with water, either by immersion or by the addition of at least 6% moisture to the fine aggregate, and permit it to stand for 24 ± 4 hours.
4. Decant excess water with care to avoid loss of fines, spread the sample on a flat, nonabsorbent surface exposed to a gently moving current of warm air, and stir frequently to cause homogeneous drying. If desired, mechanical aids such as tumbling or stirring may be used to

help achieve the saturated surface-dry condition. Continue this operation until the test specimen approaches a free-flowing condition.

5. Hold the mold firmly on a smooth, nonabsorbent surface with the large diameter down. Place a portion of the partially dried fine aggregate loosely in the mold by filling it to overflowing and heaping additional material above the top of the mold by holding it with the cupped fingers of the hand.
6. Lightly tamp the fine aggregate into the mold with 25 light drops of the tamper. Each drop should start about 5 mm above the top of surface of the aggregate. Permit the tamper to fall freely under gravitational attraction on each drop.
7. Remove loose sand from the base and lift the mold vertically. If the surface moisture is still present, the fine aggregate will retain the molded shape. If this is the case, allow the sand to dry and repeat steps 4, 5, and 6 until the fine aggregate slumps slightly indicating that it has reached a surface-dry condition.
8. Weigh 500 ± 10 g of SSD sample and record the weight; record as S.
9. Partially fill the pycnometer with water and immediately introduce into the pycnometer the SSD aggregate weighed in step 8. Fill the pycnometer with additional water to approximately 90% of the capacity. Roll, invert, and agitate the pycnometer to eliminate all air bubbles. Fill the pycnometer with water to its calibrated capacity.
10. Determine the total weight of the pycnometer, specimen, and water, and record it as C.
11. Carefully work all of the sample into a drying pan. Place in a 110 ± 10 °C oven until it dries to a constant weight. Record this weight as A.

Analysis and Results

- Bulk specific gravity $= A/(B+S-C)$

   where

   $A =$ mass of oven-dry specimen in air, g
   $B =$ mass of pycnometer filled with water, g
   $S =$ mass of the saturated surface-dry specimen, g
   $C =$ mass of pycnometer with specimen and water to the calibration mark, g

   Bulk specific gravity (SSD) $= S/(B+S-C)$
   Apparent specific gravity $= A/(B+A-C)$
   Absorption, % $= [(S-A)/A] \times 100$

Report

- Bulk specific gravity

- Bulk specific gravity SSD
- Apparent specific gravity
- Absorption

7. Bulk Unit Weight and Voids in Aggregate

ASTM Designation

ASTM C29—Bulk Density (Unit Weight) and Voids in Aggregate

Purpose

To determine the bulk unit weight and voids in aggregate in either a compacted or loose condition.

Significance and Use

The bulk density of aggregate is needed for the proportioning of portland cement concrete mixtures. The bulk density may also be used to determine the mass/volume relationships for conversions in purchase agreements. The percentage of voids between the aggregate particles can also be determined, on the basis of the obtained bulk density.

Apparatus

- Measure. Use a rigid metal watertight container with a known volume. A minimum volume of the measure is required for different nominal maximum sizes of coarse aggregate. For a 25-mm nominal maximum aggregate size, a minimum volume measure of 0.0093 m^3 is required.
- Balance, tamping rod, shovel or scoop, and a plate glass.

Test Procedure

1. Calibrate the measure as follows:
   a. Fill the measure with water at room temperature and cover with a plate glass in such a way as to eliminate bubbles and excess water.
   b. Determine the mass of the water in the measure.
   c. Measure the temperature of the water, and determine its density as shown in the table. Interpolate as necessary.

Temperature		Density	
°C	°F	kg/m³	lb/ft³
15.6	60	999.01	62.366
21.1	70	997.97	62.301
26.7	80	996.59	62.216
29.4	85	995.83	62.166

 d. Calculate the volume of the measure by dividing the mass of the water by its density.
2. Fill the measure with aggregate and compact it, either by rodding (for aggregates having nominal maximum size of 37.5 mm or less), jigging (for aggregates having a nominal maximum size of 37.5 to 125 mm), or shoveling (if specifically stipulated).
a. Rodding Procedure: Fill the measure with aggregate in three layers of approximately equal volumes. Rod each layer of aggregate with 25 strokes of the tamping rod, evenly distributed over the surface.
b. Jigging Procedure: Fill the measure with aggregate in three layers of approximately equal volumes. Compact each layer by placing the measure on a firm base, raising the opposite sides alternately about 50 mm, and allowing the measure to drop 25 times on each side.
c. Shoveling Procedure: Fill the measure to overflowing by means of a shovel or scoop, discharging the aggregate from a height not to exceed 50 mm above the top of the measure. Exercise care to avoid segregation.
3. Level the surface of the aggregate with the fingers or a straightedge. Determine the net mass of the aggregate to the nearest 0.05 kg.

Analysis and Results

$$M = \frac{G-T}{V}$$

$$\% \text{Voids} = \frac{(SW)-M}{SW} \times 100$$

where
 M = bulk unit weight of aggregate, kg/m³ (lb/ft³)
 G = weight of the aggregate plus the measure, kg(lb)
 T = weight of the measure, kg(lb)
 V = volume of the measure, m³ (ft³)
 S = bulk specific gravity (dry basis) (ASTM C127 or C128)
 W = unit weight of water, 998 kg/m³ (62.3 lb/ft³)

Report

- The bulk unit weight (or loose bulk unit weight in case of shoveling), void content, and method of compaction.

8. Slump of Freshly Mixed Portland Cement Concrete

ASTM Designation
ASTM C143—Slump of Portland Cement Concrete

Purpose
To determine the slump of freshly mixed portland cement concrete, both in the laboratory and in the field.

Significance and Use
This method measures the consistency of freshly mixed portland concrete cement (PCC). To some extent, this test indicates how easily concrete can be placed and compacted, or the workability of concrete. This test is used both in the laboratory and in the field for quality control.

Apparatus
- Mold in the form of lateral surface of frustum with a top diameter of 102 mm, bottom diameter of 203 mm, and height of 305 mm (Figure 6.3)
- Tamping rod with a length of 0.6 m, diameter of 16 mm, and rounded ends

Test Procedure
1. Mix concrete either manually or with a mechanical mixer. If a large quantity of mixed concrete exits, obtain a representative sample.
2. Dampen the mold and place it, with its larger base at the bottom, on a flat, moist, nonabsorbent rigid surface.
3. Hold the mold firmly in place by standing on the two foot pieces.
4. Immediately fill the mold in three layers, each approximately one-third of the volume of the mold. Note that one-third of the volume is equivalent to a depth of 67 mm, whereas two-thirds of the volume is equivalent to 155 mm.
5. Rod each layer 25 strokes using the tamping rod. Uniformly distribute the strokes over the

cross section of each layer. Rod the second and top layers each throughout its depth so that the strokes penetrate the underlying layer. In filling and rodding the top layer, heap the concrete above the mold before rodding is started. If the rodding operation results in subsidence of concrete below the top edge of the mold, add additional concrete to keep an excess of concrete above the top of the mold at all times.

6. After the top layer has been rodded, strike off the surface of concrete by means of a screening and rolling motion of the tamping rod.
7. Remove the mold immediately from the concrete by raising it up carefully without lateral or torsional motion. The slump test must be completed within 2.5 minutes after taking the sample.
8. Measure the slump by determining the vertical difference between the top of the mold and the displaced original center of the top of the specimen. If two consecutive tests on a sample of concrete show a falling away or a shearing off of a portion of concrete from the mass of the specimen, the concrete probably lacks the necessary plasticity and cohesiveness for the slump test to be applicable and the test results will not be valid.

Report

◆ The slump value to the nearest 5 mm

9. Unit Weight and Yield of Freshly Mixed Concrete

ASTM Designation

ASTM C138—Unit Weight, Yield, and Air Content (Gravimetric) of Concrete

Purpose

To determine the unit weight, yield, cement content, and air content of freshly mixed portland cement concrete. Yield is defined as the volume of concrete produced from a mixture of known quantities of the component materials.

Significance and Use

The unit weight value is used to calculate the volume of portland cement concrete produced from a mixture of known quantity.

Apparatus

- Measure. Use a rigid metal watertight container with a known volume. A minimum volume of the measure is required for different nominal maximum sizes of coarse aggregate. For a 25 mm nominal maximum aggregate size, a minimum volume measure of 6 liters is required.
- Balance, tamping rod, internal vibrator(optional), strike-off plate, and mallet.

Test Procedure

1. Place the freshly mixed concrete in the measure in three layers of approximately equal volume.
2. Rod each layer with 25 strokes of the tamping rod when a 0.014 m³ or smaller measure is used, otherwise use 50 strokes per layer. Distribute the strokes uniformly over the cross section of the measure. Rod the bottom layer throughout its depth, but do not forcibly strike the bottom of the measure. For the top two layers, penetrate about 25 mm into the underling layer.
3. Tap the sides of the measure smartly 10 to 15 times with the mallet to release trapped air bubbles.
4. An internal vibrator can be used instead of the tamping rod. In this case, the concrete is placed and vibrated in the measure in two approximately equal layers.
5. When the consolidation is complete, the measure must not contain a substantial excess or deficiency of concrete. A small quantity of concrete may be added to correct a deficiency.
6. After consolidation, strike off the top surface of the concrete, and finish it smoothly with the flat strike-off plate, using great care to leave the measure just level full.
7. After strike-off, clean all excess concrete from the exterior of the measure and determine the net weight of the concrete in the measure.

Analysis and Results

- $W = $ (Net weight of concrete)/(Volume of the measure)
- $Y(\text{m}^3) = W_1/W$
  $Y_f(\text{ft}^3) = W_1/W$
  $Y(\text{yd}^3) : W_1/(27W)$
- $R_y = Y/Y_d$
- $N = N_t/Y$
- $A = [(T-W)/T] \times 100$

where

W = unit weight of concrete, kg/m³ (lb/ft³)
Y = yield = volume of concrete produced per batch, m³ (yd³)
Y_f = yield = volume of concrete produced per batch, ft³
W_1 = total weight of all materials batched, kg(lb)
R_y = relative yield
Y_d = volume of concrete that the batch was designed to produce, m³ (yd³)
N = actual cement content, kg/m³ (lb/yd³)
N_t = weight of cement in the batch, kg(lb)
A = air content (percentage of voids) in the concrete
T = theoretical unit weight of concrete computed on an air-free basis, kg/m³ (lb/ft³)

Report

♦ The unit weight, yield, relative yield, actual cement content, and air content

10. Air Content of Freshly Mixed Concrete by Pressure Method

ASTM Designation

ASTM C231—Air Content of Freshly Mixed Concrete by Pressure Method

Purpose

To determine the air content of freshly mixed portland cement concrete by the pressure method.

Significance and Use

Air content plays an important role in workability of freshly mixed concrete and the strength and durability of hardened concrete. The air content of freshly mixed concrete is needed for the proper proportioning of the concrete mix.

Apparatus

♦ Air meter type B, consisting of a measuring bowl of a capacity at least 0.006 m³ and cover assembly fitted with air valves, air bleeder valves, petcocks, and suitable hand pump. The air meter must be frequently calibrated according to ASTM C231 procedure to ensure

- Miscellaneous items, including trowel, tamping rod, mallet, and strike-off bar.

Test Procedure

1. Place a representative sample of the plastic concrete in the measuring bowl in three equal layers.
2. Consolidate each layer of concrete by 25 strokes of the tamping rod, evenly distributed over the cross section.
3. After rodding each layer, tap the sides of the measuring bowl 10 to 15 times with the mallet to remove any voids.
4. Strike off the top surface by sliding the bar across the top rim with a sawing motion.
5. Thoroughly clean the flanges and cover the assembly to obtain a pressure-tight seal.
6. Using a rubber syringe, inject water through one petcock until water emerges from the opposite petcock.
7. Jar the meter gently until all air is expelled from the same petcock.
8. Pump air into the air chamber until the gauge indicator is on the initial line.
9. Open the air valve between the air chamber and the measuring bowl.
10. Tap the sides of the measuring bowl sharply, and lightly tap the pressure gauge.
11. Read the percentage of air content on the dial gauge.
12. Determine the aggregate correction factor according to ASTM C231, and subtract it from the reading obtained in step 11.

Report

- The air content and the method used (pressure method).

11. Air Content of Freshly Mixed Concrete by Volumetric Method

ASTM Designation

ASTM C173—Air Content of Freshly Mixed Concrete by Volumetric Method

Purpose

To determine the air content of freshly mixed portland cement concrete by the volumetric method.

Significance and Use

Air content plays an important role in the workability of freshly mixed concrete and the strength and durability of hardened concrete. The air content of freshly mixed concrete is needed for the proper proportioning of the concrete mix.

Apparatus

- Air meter, consisting of a bowl and a top section, as shown in Figure 6.4
- The bowl has a diameter of 1 to 1.25 the height and a minimum capacity of 0.002 m^3 (0.075 ft^3). The capacity of the top section is 1.2 times the capacity of the bowl.
- Miscellaneous items including funnel, tamping rod, strike-off bar, measuring cup, syringe, pouring vessel, trowel, scoop, isopropyl alcohol, and mallet.

Calibration

1. The volume of the bowl must be calibrated by accurately weighing the amount of water required to fill it at room temperature and dividing this weight by the unit weight of water at the same temperature.
2. The accuracy of the graduation on the neck of the top section and the volume of the measuring cup must be calibrated according to ASTM C173.

Test Procedure

1. Fill the bowl with freshly mixed concrete in three layers of equal depth.
2. Rod each layer 25 times with the tamping rod.
3. After each layer is rodded 10 to 15 times, tap the sides of the measuring bowl with the mallet to release air bubbles.
4. After placement of the third layer of concrete, strike off the excess concrete with the strike-off bar until the surface is flush with the top of the bowl. Wipe the flanges of the bowl clean.
5. Clamp the top section on the bowl, insert the funnel, and add water until it appears in the neck. Remove the funnel and adjust the water level, using the rubber syringe, until the bottom of the meniscus is leveled with the zero mark. Attach and tighten the screw cap.
6. Invert and agitate the unit many times until the concrete settles free from the base.
7. When all the air rises to the top of the apparatus, remove the screw cap. Add, in 1-cup increments using the syringe, sufficient isopropyl alcohol to dispel the foamy mass on the surface of the water. Note that the capacity of the cup is equivalent to 1.0% of the volume of the bowl.

8. Make a direct reading of the liquid in the neck to the bottom of the meniscus to the nearest 0.1%.
9. The percent air content is calculated as the reading in step 8 plus the amount of alcohol used.

Report

◆ The air content and the method used (volumetric method).

12. Making and Curing Concrete Cylinders and Beams

ASTM Designation

ASTM C31—Making and Curing Concrete Test Specimens

Purpose

To determine how to make and cure concrete cylindrical and beam specimens.

Significance and Use

This practice provides standardized requirements for making and curing portland cement concrete test specimens. Specimens can be used to determine strength for mix design, quality control, and quality assurance.

Apparatus

◆ Cylindrical molds made of steel or another nonabsorbent and nonreactive material. The standard specimen size used to determine the compressive strength of concrete is 152 mm diameter by 304 mm high for a maximum aggregate size up to 50 mm. Smaller specimens, such as 102 mm diameter by 203 mm high, are sometimes used, but they are not ASTM standards.

◆ Beam molds made of steel or another nonabsorbent, nonreactive material. Several mold dimensions can be used to make beam specimens with a square cross section and a span three times the depth. The standard ASTM inside mold dimensions are 152 mm × 152 mm in cross section and a length of not less than 508 mm, for a maximum aggregate size up to 50 mm.

◆ Tamping rod with a length of 0.6 m, diameter of 16 mm, and rounded ends.

◆ Moist cabinet or room with not less than 95% relative humidity and 23 ± 1.7 °C (73 ± 3

°F) temperature or a large container filled with lime saturated water for curing
- Miscellaneous items including vibrator(optional), scoop, and trowel.

Test Procedure

1. Weigh the required amount of coarse aggregate, fine aggregate, portland cement, and water.
2. Mix the materials in the mixer for 3 to 5 min. If an admixture is used, it should be mixed with water before being added to the other materials.
3. Check slump, air content, and temperature of concrete.
4. For cylindrical specimens, place concrete into the mold using a scoop or trowel. Fill the cylinder in three equal layers, and rod each layer 25 times. Tap the outside of the cylinder 10 to 15 times after each layer is rodded. Strike off the top and smooth the surface. Vibrators can also be used to consolidate the concrete instead of rodding. Vibration is optional if the slump is between 25 mm to 75 mm and is required if the slump is less than 25 mm.
5. For beam specimens, grease the sides of the mold and fill the molds with concrete in two layers. Consolidate the concrete by either tamping each layer 60 times until uniformly distributed throughout or by vibrating. After consolidation, finish the surface by striking off the surface and smoothing.
6. Cover the mold with wet cloth to prevent evaporation.
7. Remove the molds after 16 hours to 32 hours.
8. Cure the specimen in a moist cabinet or room at a relative humidity of not less than 95% and a temperature of 23 ± 1.7 °C (73 ± 3 °F) or by submersion in lime-saturated water at the same temperature.

Precautions

1. Segregation must be avoided. Over vibration may cause segregation.
2. In placing the final layer, the operator should attempt to add an amount of concrete that will exactly fill the mold after compaction. Do not add nonrepresentative concrete to an underfilled mold.
3. Avoid overfilling by more than 6 mm.

Report

- Record mix design weights, slump, temperature of the mix, and air content.
- Specimen type, number of specimens, dimensions, and any deviations from the standard preparation procedure.
- Include this information with the report on the strength of the concrete.

13. Flexural Strength of Concrete

ASTM Designation
ASTM C78—Flexural Strength of Concrete (Using Simple Beam with Third-Point Loading)

Purpose
To determine the flexural strength of portland cement concrete by using a simple beam with third-point loading.

Significance and Use
The flexural strength of concrete is a measure of concrete quality.

Apparatus
- Loading machine capable of applying loads at a uniform rate
- Loading device capable of applying load configuration as shown in Figure 6.11. Forces applied to the beam shall be perpendicular to the face of the specimen and applied without eccentricity.

Test Specimens
- The standard ASTM specimen dimensions are 152 mm × 152 mm in cross section and a length of not less than 508 mm for a maximum aggregate size up to 50 mm.
- Sides of the specimen should be at right angles to its top and bottom. All surfaces in contact with load-applying and support blocks should be smooth and free of scars, indentations, holes, or inscribed identifications.

Test Procedure
1. Turn the test specimen on its side, with respect to its position as molded, and center it on the bearing blocks.
2. Center the loading system in relation to the applied force. Bring the load-applying blocks in contact with the surface of the specimen at the third points between the supports.
3. If full contact is not obtained at no load between the specimen and the load-applying blocks and the supports so that there is a 25 mm or larger gap in excess of 0.1 mm, grind or cap the contact surfaces of the specimen, or shim with leather strips.

4. Apply the load rapidly up to approximately 50% of the breaking load. Thereafter, apply the load continuously at a rate that constantly increases the extreme fiber stress between 860 kPa and 1210 kPa/min until rupture occurs.

Analysis and Results

- Take three measurements across each dimension (one at each edge and at the center) to the nearest 1.3 mm to determine the average width, average depth, and line of fracture location of the specimens at the section of fracture.
- If the fracture initiates in the tension surfaca within the middle third of the span length, calculate the modulus of rupture as follows:

$$R = \frac{Mc}{I} = \frac{PL}{bd^2}$$

where

R = modulus of rupture, MPa(psi)

M = maximum bending moment, N · mm(lb. in.)

$c = \frac{d}{2}$, mm(in.)

I = moment of inertia

$= \frac{bh^3}{12}$, mm^4(in^4.)

P = maximum load, N(lb)

L = span length, mm(in.)

b = average width, mm(in.)

d = average depth, mm(in.)

- If the fracture occurs in the tension surface outside the middle third of the span length, by not more than 5% of the span length, calculate the modulus of rupture as follows:

$$R = \frac{3Pa}{bd^2}$$

where

a = average distance between line of fracture and the nearest support on the tension surface of the beam in millimeters

- If the fracture occurs in the tension surface outside the middle third of the span length, by more than 5% of the span length, discard the results of the test.

Report

- Specimen identification number.

- Average width.
- Average depth.
- Span length.
- Maximum applied load.
- Modulus of rupture to the nearest 0.03 MPa.
- Curing history and apparent moisture condition at time of testing.
- If specimens were capped, ground, or if leather shims were used.
- Defects in specimens.
- Age of specimens.

14. Penetration Test of Asphalt Cement

ASTM Designation

ASTM D5—Penetration of Bituminous Materials

Purpose

To determine the penetration of semisolid and solid bituminous materials.

Significance and Use

The penetration test is used as a measure of consistency. High values of penetration indicate soft consistency.

Apparatus

- Penetration apparatus and needle (Figures A.13)
- Sample container, water bath, transfer dish, timing device for hand-operated penetrometers, and thermoneters

Test Procedure

1. Heat the asphalt binder sample until it has become fluid enough to pour.
2. Pour the sample into the sample container and let it cool for at least 1 hour.
3. Place the sample, together with the transfer dish in the water bath at a temperature of 25 °C (77 °F) for 1 hour to 2 hours.
4. Clean and dry the needle with a clean cloth, and insert the needle into the penetrometer. Unless otherwise specified, place the 50-g mass above the needle, making the total moving load 100 g.

Figure A.13 Apparatus for penetration test of asphalt binder.

5. Place the sample container in the transfer dish, cover the container completely with water from the constant temperature bath, and place the transfer dish on the stand of the penetrometer.
6. Position the needle by slowly lowering it until its tip just makes contact with the surface of the sample. This is accomplished by bringing the actual needle tip into contact with its image reflected by the surface of the sample from a properly placed source of light.
7. Quickly release the needle holder for the specified period of time (5 seconds) and adjust the instrument to measure the distance penetrated in tenths of a millimeter.
8. Make at least three determinations at points on the surface of the sample not less than 10 mm from the side of the container and not less than 10 mm apart.

Report

◆ The average of the three penetration values to the nearest whole unit

15. Absolute Viscosity Test of Asphalt

ASTM Designation
ASTM D2171—Viscosity of Asphalts by Vacuum Capillary Viscometer

Purpose
To determine the absolute viscosity of asphalt by vacuum capillary viscometer at 60 °C (140 °F).

Significance and Use
The viscosity at 60 °C (140 °F) characterizes flow behavior and may be used for specification requirements for cutbacks and asphalt cements.

Apparatus
- Viscometers such as the Cannon-Manning vacuum viscometer or modified Koppers vacuum viscometer
- Bath, with provisions for visibility of the viscometer and the thermometer
- Thermometers, vacuum system, and timing device

Test Procedure
1. Maintain the bath at a temperature of 60 °C (140 °F).
2. Select a clean, dry viscometer that will give a flow time greater than 60 seconds, and preheat to 135 °C (275 °F).
3. Charge the viscometer by pouring the prepared asphalt sample to the fill line.
4. Place the charged viscometer in an oven or bath maintained at 135 °C for a period of 10 min, to allow large air bubbles to escape.
5. Remove the viscometer from the oven or bath, and within 5 min, insert the viscometer in a holder, and position the viscometer vertically in the bath so that the uppermost timing mark is at least 20 mm below the surface of the bath liquid.
6. Establish a 300-mm Hg vacuum in the vacuum system, and connect the vacuum system to the viscometer.
7. After the viscometer has been in the bath for 30 min, start the flow of asphalt in the viscometer by opening the toggle valve or stopcock in the line leading to the vacuum system.

8. Measure to within 0.1 second the time required for the leading edge of the meniscus to pass between successive pairs of timing marks. Report the first flow time that exceeds 60 seconds between a pair of timing marks, and identify the pair of timing marks.

Analysis and Results

◆ Select the calibration factor that corresponds to the pair of timing marks. Calculate and report the viscosity to three significant figures formula
$$P = Kt$$
where
P = absolute viscosity, Poises
K = selected calibration factor, Poises/s
t = flow time, s

Report

◆ Absolute viscosity.
◆ Test temperature and vacuum.

16. Preparing and Determining the Density of Hot-Mix Asphalt (HMA) Specimens by Means of the Superpave Gyratory Compactor

AASHTO Designation

AASHTO T 312-03—Preparing and Determining the Density of Hot-Mix Asphalt (HMA) Specimens by Means of the Superpave Gyratory Compactor

Purpose

To prepare asphalt concrete specimens to densities achieved under actual pavement climate and loading conditions, or as needed for laboratory testing.

Significance and Use

The Superpave gyratory compactor is capable of accommodating large aggregates. Furthermore, this device affords a measure of compaction ability that permits potential tender mixture behavior and similar compaction problems to be identified. This method of compaction is used for the Superpave volumetric mix design of asphalt concrete mixture, and for field quality control during the construction of HMA pavements.

Apparatus

- Reaction frame, rotating base, and motor (Figure 8.8)
- Loading system, loading ram, and pressure gauge
- Height measuring and recording system
- Mold and base plate

Procedure

1. Specimens must be mixed and compacted under equiviscous temperature conditions of 0.170 ± 0.02 Pa·s and 0.280 ± 0.03 Pa·s, respectively. Mixing is accomplished by a mechanical mixer.
2. After mixing, loose test specimens are subjected to 2 hours of short-term aging in a forced draft oven at the compaction temperature which corresponds to a binder viscosity of 0.28 ± 0.03 Pa·s. During this period, loose mix specimens are required to be spread into a thickness resulting in 21 kg to 22 kg per cubic meter. The sample is stirred after 1 hour to ensure uniform aging.
3. Place the compaction molds and base plates in an oven at the compaction temperature for at least 30 min to 45 min prior to use.
4. If specimens are to be used for volumetric determinations only, use sufficient mix to arrive at a specimen 150 mm in diameter by approximately 115 mm high. This requires approximately 4500 g of aggregates. If needed, specimens with other heights can also be prepared for further performance testing.
5. Turn on the power to the compactor. Set the vertical pressure to 600 kPa.
6. Set the gyration counter to zero, and set it to stop when the desired number of gyrations is achieved. Three gyrations are of interest: design number of gyrations(N_d), initial number of gyrations(N_i), and maximum number of gyrations(N_m). The design number of gyrations is a function of the climate in which the mix will be placed and the traffic level it will withstand.
7. After the base plate is placed, place a paper disk on top of the plate and charge the mold with the short-term aged mix in a single lift. The top of the uncompacted specimen should be slightly rounded. Place a paper disk on top of the mixture. Depending on the model of the compactor it may be necessary to place a top plate on the mix.
8. Place the mold in the compactor and center it under the ram. Lower the ram until it contacts the mixture and the resisting pressure is 600 kPa.
9. Apply the angle of gyration of 1.25, and begin compaction.
10. When the desired number of gyrations has been reached, the compactor should automatical-

ly cease. After the angle and pressure are released, remove the mold containing the compacted specimen.
11. Print the results of specimen height versus number of gyrations.
12. After a suitable cooling period, extrude the specimen from the mold and mark it.
13. Measure the bulk specific gravity of the test specimens according to ASTM D2726 procedure.

Analysis and Results

- Compute the estimated bulk specific gravity for each desired gyration from the formula
  Estimated bulk specific gravity = net weight of the specimen/$(\pi d^2 h/4)$
  where
  d = mold diameter (150 mm)
  h = specimen height corresponding to the desired number of gyrations
- Compute the correction factor C as follows:
  $$C = \frac{\text{measured bulk specific gravity}}{\text{estimated bulk specific gravity at } N_m}$$
- Compute the corrected bulk specific gravity by multiplying each estimated bulk specific gravity by the correction factor.
- Compute the percentage of the corrected bulk specific gravity relative to the maximum theoretical specific gravity ($\%G_{mm}$).
- Draw a graph of the logarithm of the number of gyrations versus the $\%G_{mm}$ (densification curve) for each specimen.

Report

- Mixture ingredients, source, and relevant information
- Densification table
- Densification curve

17. Preparation of Asphalt Concrete Specimens Using the Marshall Compactor

ASTM Designation

ASTM D1559—Resistance to Plastic Flow of Bituminous Mixtures Using Marshall Apparatus

Purpose

To prepare asphaltic concrete specimens using the Marshall hammer.

Significance and Use

This method is used to design the mix using the Marshall procedure and to measure its properties.

Apparatus

- Either mechanical or manual compaction hammer with 4.5-kg weight and 0.48-m drop height can be used (Figure 8.9)
- Molds with 102-mm (4-in.) inside diameter and 75 mm (3 in.) high, base plates, and collars
- Compaction pedestal, specimen extruder, and miscellaneous items, such as mold holder, spatula, pans, and oven.

Procedure

1. Determine the mixing and compaction temperatures so that the kinematic viscosities of the binder are 170 ± 20 cSt and 280 ± 30 cSt, respectively.
2. Separate all the required sizes of aggregates and oven dry.
3. Weigh 1200 g batches so that gradation would satisfy the midpoint of the specification band. Either a state specification or ASTM D3515 can be followed.
4. Place both asphalt binder and aggregate in the oven until they reach the mixing temperature (approximately 150 °C or 300 °F).
5. Add the asphalt to the aggregate in the specified amount.
6. Using the mechanical mixer, or manually, mix the aggregates and asphalt thoroughly. Some agencies require curing the mix as specified in AASHTO R30.
7. Place a release paper inside the mold. Then, place the entire batch of asphaltic concrete in the mold heated to the compaction temperature, and spade with a heated spatula 15 times around the perimeter and 10 times in the middle.
8. Put in place a collar and a release paper, and put the mold on the pedestal. Clamp the mold with the mold holder, and apply the required number of blows using either a manual hammer or a mechanical hammer. The typical number of blows on each side is either 50 or 75, depending on the expected traffic volume on the road where the mix is intended to be used. Invert the mold and apply the same number of blows on the other face.
9. After cooling to room temperature, extrude the specimen using an extruding device. Cooling can be accelerated by placing the mold and the specimen in a plastic bag and subjecting them

to cold water. Remove the paper disk, as shown in Figure A.14.

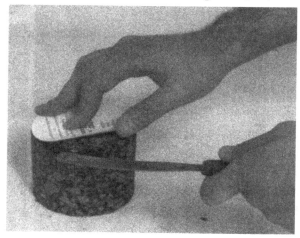

Figure A.14 Removing paper disk from a Marshall plug.

Report

- Mixture ingredients, source, and relevant information
- Number of blows on each side of the specimen

18. Bulk Specific Gravity of Compacted Bituminious Mixtures

ASTM Designation

ASTM D2726—Bulk Specific Gravity of Compacted Bituminous Mixtures

Purpose

To determine the bulk specific gravity of compacted asphalt mixture specimens.

Significance and Use

The results of this test are used for voids analysis of the compacted asphalt mix.

Test Specimens

Laboratory-molded bituminous mixtures or cores drilled from bituminous pavements can be used.

Apparatus

- Balance equipped with suitable suspension and holder to permit weighing the specimen while it is suspended from the balance
- Water bath

Test Procedure

1. Weigh the specimen in air and record it as A.
2. Immerse the specimen in water at 25 ± 1 °C (77 ± 2 °F) while it is suspended from the balance for 3 min to 5 min, and record the immersed weight as C.
3. Remove the specimen from the water, surface dry by blotting with a damp towel, determine the surface dry weight, and record it as B.

Analysis and Results

- Calculate the bulk specific gravity as

$$\text{Bulk Specific Gravity} = \frac{A}{(B-C)}$$

where

A = mass of specimen in air, g
B = mass of surface-dry specimen, g
C = mass of specimens in water, g

Report

- Report the value of specific gravity to three decimal places.

19. Marshall Stability and Flow of Asphalt Concrete

ASTM Designation

ASTM D1559—Resistance to Plastic Flow of Bituminous Mixtures Using Marshall Apparatus

Purpose

To determine the Marshall stability and flow values of asphalt concrete.

Significance and Use

This test method is used in the laboratory mix design of bituminous mixtures according to the Marshall procedure. The test results are also used to characterize asphalt mixtures.

Apparatus

- Testing machine producing a uniform vertical movement of 50.8 mm minute, as shown in Figure 8.19
- Breaking heads having an inside radius of curvature of 50.8 mm
- Load cell or ring dynamometer, strip chart recorder or flow meter, water bath, and rubber gloves

Test Procedure

1. Bring the specimen prepared in experiment number 25 to a temperature of 60 ℃ (140 °F) by immersing it in a water bath 30 min to 40 min or by placing it in the oven for 2 hours.
2. Remove the specimen from the water bath then "lightly" dry (or oven) and place it in the lower segment of the breaking head. Place the upper segment of the breaking head on the specimen, and place the complete assembly in position on the testing machine.
3. Prepare the strip chart recorder or place the flowmeter (where used) in position over one of the guide rods, and adjust the flowmeter to zero while holding the sleeve firmly against the upper segment of the breaking head.
4. Apply the load to the specimen by means of the constant rate of movement of 50.8 mm/min until the maximum load is reached and the load decreases (Figure A.15). The elapsed time for the test from removal of the test specimen from the water bath (or oven) to the maximum load determination should not exceed 30 sec.
5. From the chart recorder, record the Marshall stability (maximum load) and the Marshall flow (deformation when the maximum load begins to decrease in units of 0.25 mm or hundredths of an inch). In some machines the maximum load and the flow values are read from the ring dynamometer and the flowmeter, respectively.
6. If the specimen height is other than 63.5 mm, multiply the stability value by a correction factor (Table 8.5) (ASTM D1559).

Report

- Specimen identification and type (laboratory prepared or core)
- Average Marshall stability of at least three replicate specimens, corrected when required, kN(lb)

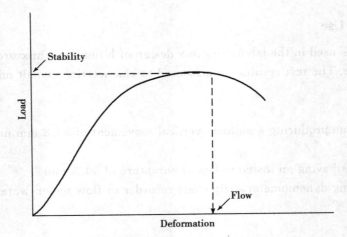

Figure A.15 Typical plot of load versus deformation during the Marshall stability test showing the stability and flow.

- Average Marshall flow of at least three replicate specimens

20. Bending and Compression Tests of Wood

ASTM Designation

ASTM D143—Standard Methods of Testing Wood

Purpose

To determine modulus of rupture and compressive strength of wood by testing clear specimens.

Significance and Use

These tests provide data for comparing the mechanical properties of various species and data for the establishment of strength functions. The tests also provide data to determine the influence of such factors as density, locality of growth, change of properties with seasoning or treatment with chemicals, and changes from sapwood to heartwood on the mechanical properties.

Static Bending Test

Specimens 50 mm×50 mm×760 mm are used for the primary method, and 25 mm×25 mm×410 mm for the secondary method. For the loading span and supports, use center loading and a span length of 710 mm for the primary method and 360 mm for the secondary method.

Apparatus

- Testing machine of the controlled deformation type
- Bearing blocks for applying the load.

Test Procedure

1. Place the specimen so that the load will be applied at the center of the span. The load is applied through the bearing block to the tangential surface nearest the pith(Figure A.16).
2. Apply the load continuously throughout the test at a rate of motion of 2.5 mm/min for primary method specimens, and at a rate of 1.3 mm/min for secondary method specimens.
3. Record the load-deflection curve up to or beyond the maximum load. Continue recording up to a 150 mm deflection, or until the specimen fails to support a load of 890 N(200 lb) for the primary method specimens, and up to a 76 mm deflection or until the specimen fails to support a load of 220 N for secondary method specimens.
4. Within the proportional limit, take deflection readings to the nearest 0.02 mm. After the proportional limit is reached, the deflection is read by means of the dial gauge until it reaches the limit of its capacity, normally 25 mm. Where deflections beyond 25 mm are encountered, the deflections may be read by means of the scale mounted on the moving head.
5. Read the load and deflection of the first failure, the maximum load, and points of sudden change, and plot the load-deformation data.

Figure A.16 Wood specimen during static bending test.

Analysis and Results

- Calculate the modulus of rupture as

$$R = \frac{Mc}{I}$$

where

R = modulus of rupture, MPa(psi)

M = bending moment = $PL/4$

P = maximum load, N(lb)
L = span length, mm(in.)
C = distance from neutral axis to edge of sample = $\frac{1}{2}h$
I = moment of inertia = $bh^3/12$
b = average width, mm(in.)
h = average depth, mm(in.)

- Static bending (flexural) failures are classified in accordance with the appearance of the fractured surface and the manner in which the failure develops, as shown in Figure A.17. The fracture failure may be roughly divided into brash and fibrous, the term brash indicating abrupt failure and fibrous indicating a fracture showing splinters.

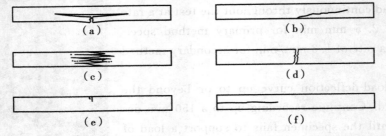

Figure A.17 Types of failure in static bending: (a) simple tension (side view), (b) cross-grain tension (side view), (c) splintering tension (view of tension surface), (d) brash tension (view of tension surface), (e) compression (side view), and (f) horizontal shear (side view).

Report

- Specimen identification, dimensions, span length, and other relevant information, such as moisture content
- Load-deflection plot
- Modulus of rupture
- Failure condition

Compression Parallel to the Grain Test

Specimens 50 mm × 50 mm × 200 mm are used for the primary method, and 25 mm × 25 mm × 100 mm for the secondary method. Be careful that the end grain surfaces are parallel to each other and at right angles to the longitudinal axis when preparing the specimens.

Apparatus

A controlled deformation machine is required. At least one platen of the testing machine is equipped with a spherical bearing to obtain uniform distribution of load over the ends of the specimen.

Test Procedure

1. Place the specimen perpendicularly on the crosshead of the machine, as shown in Figures 9.4(c) and 9.5.
2. Apply the load continuously throughout the test until failure at a rate of motion of 0.003 mm/mm of the nominal specimen length per minute.
3. Measure the deformation over a central gauge length not exceeding 150 mm for primary method specimens, and 50 mm for secondary method specimens. Load-compression readings should be continued until the proportional limit is well passed, as indicated by the curve.

Analysis and Results

- Plot the load versus deflection diagram.
- Determine the modulus of elasticity as the slope of the straight portion of the stress-strain curve.
- Classify the compression failure in accordance with the appearance of the fractured surface (Figure 9.6). In case two or more kinds of failures develop, describe all fractured surfaces in the order of their occurrence; for example, shearing followed by brooming.

Report

- Specimen identification, dimensions, and other relevant information, such as moisture content.
- Load-deformation plot.
- Modulus of elasticity.
- Failure condition.

Compression Perpendicular to Grain Test

Specimens 50 mm × 50 mm × 150 mm are used.

Apparatus

- Controlled deformation machine
- Metal bearing plate 50 mm wide

Test Procedure

1. Position the specimen on the crosshead of the machine, as illustrated in Figure 9.4(d).
2. Apply the load through a metal bearing plate 50 mm wide, placed across the upper surface of the specimen at equal distances from the ends and at right angles to the length. Measure the actual width of the bearing plate. The specimens are to be placed so that the load will be applied through the bearing plate to a radial surface. The load is to be applied continuously throughout the test at a rate of 0.305 mm/min (0.012 in./min).
3. Take load and deformation readings up to 2.5 mm compression, after which discontinue the test. Measure the compression between the loading surfaces.

Analysis and Results

♦ Plot the load versus deflection diagram.
♦ Determine the modulus of elasticity as the slope of the straight portion of the stress-strain curve.

Report

♦ Specimen identification, dimensions, and other relevant information such as moisture content
♦ Load-deformation plot
♦ Modulus of elasticity

21. Tensile Properties of Plastics

ASTM Designation

ASTM D638M—Tensile Properties of Plastics

Purpose

To determine the tensile properties of unreinforced and reinforced plastic materials, including composites.

Significance and Use

This test method is designed to produce tensile property data for the control and specification of plastic materials. These data are also useful in characterizing the quality of the material and for research and development.

Apparatus

Testing machine, grips, load indicator, and extension indicator

Test Specimens

Specimens with an overall length of 150 mm, gauge length of 50 mm, width of narrow section of 10 mm, and thickness of 4 mm are used (Figure A.18) (Type M-I specimens, ASTM D638M). For isotropic materials, at least five specimens are tested.

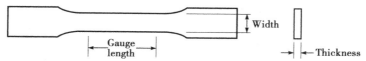

Figure A.18 Plastic specimen for tension test.

Test Procedure

1. Measure the width and thickness of specimens with a suitable micrometer to the nearest 0.02 mm at several points along their narrow sections.
2. Condition the test specimens at 23 ± 2 ℃ and $50\pm5\%$ relative humidity for not less than 40 hours prior to the test.
3. Place the specimen in the grips of the testing machine, taking care to align the long axis of the specimen and the grips with an imaginary line joining the points of attachment of the grips to the machine.
4. Attach the extension indicator (Figure A.19). When the modulus is required, the extension indicator must continuously record the distance the specimen is stretched (elongated) within the gauge length, as a function of the load through the initial (linear) portion of the load-elongation curve.
5. Set the speed of testing at a rate of travel of the moving head of 5 mm/min and start the machine (Figure A.20).
6. Record the load-extension curve of the specimen.
7. Record the load and extension at the yield point (if one exists) and the load and extension at the moment of rupture.

Analysis and Results

- Tensile strength is
$$\sigma = P_{max}/A_o$$
where

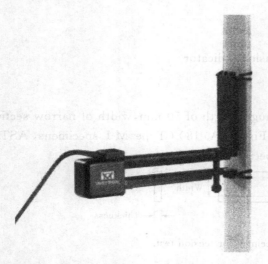

Figure A.19   Extensometer on plastic specimen.

Figure A.20   Tension test on a plastic specimen.

σ = tensile strength, MPa(psi)

P_{max} = maximum load carried by the specimen during the tension test, N(lb)

A_o = original minimum cross-sectional area of the specimen, mm² (in.²)

- Percent elongation

   If the specimen gives a yield load that is larger than the load at break, calculate the percent elongation at yield. Otherwise, calculate the percent elongation at break. Do this by reading the extension (change in gauge length) at the moment the applicable load is reached. Divide that extension by the original gauge length and multiply by 100.

- Modulus of elasticity

   Calculate the modulus of elasticity by extending the initial linear portion of the load-extension curve and by dividing the difference in stress of any segment of section on this straight line by the corresponding difference in strain. Compute all elastic modulus values using the average initial cross-sectional area of the test specimens in the calculations.

Report

- Complete identification of the material tested
- Method of preparing test specimen, type of test specimen and dimensions, and speed of testing
- Tensile strength, percent elongation, and modulus of elasticity.

GLOSSARY

A

abrasion 磨损
accelerators 速凝剂
adhesive 胶粘剂
admixtures 矿物掺合料
affinity for asphalt 沥青粘结力
air entrainment agents 引气剂
alkali-aggregate reactivity 碱-集料反应
aliphatic 脂肪质的,脂肪族的
angular 有角的,尖的;倾斜的,多角的
anionic emulsion 阴离子感光乳液
area defects 面缺陷
arithmetic mean 算术平均值
asbestos 石棉
asphalt cutback 溶于石油(馏出物)的沥青

B

batching concrete 混凝土投料
bark pocket 皮囊,夹皮
bauxite 矾土,铁铝氧石
Blaine fineness 布莱恩细度
blunder 误差
bonding 连[搭,焊,胶,粘]接,粘结剂
bone dry 完全干燥
bound water （物理或化学）结合水,束缚水
brake forming 弯曲成型
bulk specific gravity 体积密度,比重

C

Calcium silicate 硅酸钙
Calcium-silicate-h·rate(C-S-H) 水化硅酸钙
Cambium [植]形成层,新生组织
capillary voids 毛细孔隙
Carbon-content steels 碳钢
cement soundness 水泥安定性
central plant recycling 中枢装置循环
central tendency 集中趋势
central-mixed concrete 集中拌和混凝土
Chase air indicator test 蔡斯空气含量指示试验
component fractionation schemes 组分分馏系统
consistency 浓度,稠度,密度,一致性,连贯性
coating 涂料,涂层
compression parallel to the grain test 顺纹抗压试验
compression perpendicular to the grain test 横纹抗压试验
controlled density fill 密度控制性填充
conventional concrete 普通混凝土

covalent bonds 共价键
cutbacks 稀释(产物)

D

deleterious substance 有毒(害)物质
depositing concrete 浇注混凝土
dynamic shear rheometer 动力剪切流变仪

E

edge dislocation 刃型位错
emulsified asphalts 乳化沥青
emulsifying agents 乳化剂
entraining air 引气

F

facing bricks 饰面砖
false set 假凝,闪凝
fatigue cracking 疲劳裂缝
fiber-reinforced concrete 纤维增强混凝土
fineness modulus 细度模数
finishing concrete 饰面混凝土
freeze and thaw test 冻融试验

G

galvanic corrosion 电化学腐蚀;电偶腐蚀
gap-graded 间断级配;非连续级配
general yielding 总变形
GGBF-slag 粒化高炉矿渣
Graphite 石墨
gravel 砂砾
grout 灌浆

H

heavyweight concrete 重混凝土
hemicelluloses 半纤维素
hexagonal close pack structure 六方密堆结构
highly-ordered polymer 高分子聚合物
h·rophilic 亲水的,吸水的

I

I-beams 工字形梁
igneous rocks 火成岩
immersion 浸没
incoherent boundary 非共格边界
inorganic solids 无机固体
in-place recycling 原位循环
insulating blankets 绝缘垫
ionic bonds 离子键
iron-carbide 碳化铁,碳铁化物
isostrain condition 等应变条件
isotactic structure 同位立体结构
isotopes 同位素

K

killed steels 镇静钢,脱氧钢
kinematic viscosity test procedure 动粘度试验方法

L

laminated strand lumber 层叠木片胶合木
laminated veneer lumber 单板层积材,细木工板
lattice defects 晶格缺陷
lean fill 贫填充
lightweight synthetic aggregates 轻质人造集料
lignin 木质素
lithium-based admixtures 锂基掺合料
load cells 荷载单元
low-relaxation steels 低松弛钢

M

macroscopic composite 宏观复合材料
maximum density gradation 最大密度梯度
manufacture aggregate 人工集料
masonry cement 砌筑水泥
medium-curing(MC)cutback 慢凝逆转
metamorphic rocks 变质岩
Mig welding 米格(Mig)焊接法
mild steel 低碳钢

modular bricks 砖模数
moisture correction 湿度(含水量)校正

N

natural pozzolans 天然火山灰
Newtonian element 牛顿单元
Newtonian fluids 牛顿流体
noncontact extensometer 非接触式伸长计
Non-modular bricks 不以模数为基数的砖
Normal distribution 正态分布

O

offset yield stress 条件屈服极限
one-sized graded aggregates 单粒级骨料
open hearth furnace 平炉渣
open-graded 开放式级配
organic solids 有机固体
oriented strand board 定向结构刨花板,定向条板
oriented strand lumber 杉木积成材

P

parallel strand lumber 平行条板
parraffinic 石蜡基的
partially soluble materials 部分溶解材料
plain Portland cement concrete 素混凝土
Portland-pozzolan cement 火山灰水泥
pressure aging vessel 压力老化容器
pretensioned joints 预张拉接头
primary bonds 主键
proving ring 测力环,试验环
pultrusion 挤拉成型
pycnometer 比重瓶

R

raised grain 粗糙表面,波纹纹理
rapid curing(RC)cutbacks 快干沥青稀释油
raveling 被拆散的东西
rebound hammer 回弹仪
reinforced Portland cement concrete 钢筋混凝土

resilient modulus test 回弹模量试验
retarder 缓凝剂
rheological model 流变学模型
rivet fasterens 铆钉固定
rutting 车辙

S

sacrificial primers 衬底
saturated rings 饱和环
secondary bonds 次级键,副键
sedimentary rocks 沉积岩
semiguided bend test 半引导弯曲试验
shotcrete 喷射混凝土
shrink-mixed concrete 混合收缩混凝土
slip-critical joints 临界滑移点
snug-tightened joints 紧固高强螺栓
space lattice 空间点阵
spreading concrete 摊铺混凝土
steel rabars 螺纹钢
stripping 脱模,拆模
soundness 坚固性
specific gravity 密度
superplastic forming 超塑性成型

T

thermoplastics 热塑性塑料
thermosets 热固性,热固树脂
trapped air 捕获的空气
toughness 韧性

U

unit cell 晶胞,晶格
unit rate 单位比率
unshrinked fill 不收缩填充物
underlying materials 基础材料

V

valence electrons 价电子,价电子数
veneer-based materials 胶合板基材料

vibration tables 振动台
vibratory rollers 振动压路机
vicat test 维卡试验
viscoelastic materials 粘弹性材料
viscous flow 粘性流，粘滞流

vitreous ceramics 玻璃质陶瓷

W

wet covering 湿覆盖物
wire fabrics 钢丝围栅